# LEARNING RESOURCE MANUAL to ACCOMPANY
## FUNDAMENTALS OF
# PHYSIOLOGY
### A HUMAN PERSPECTIVE

### SECOND EDITION

**LAURALEE SHERWOOD**

DEPARTMENT OF PHYSIOLOGY
SCHOOL OF MEDICINE
WEST VIRGINIA UNIVERSITY

**DAVID P. SHEPHERD**

DEPARTMENT OF BIOLOGICAL SCIENCES
SOUTHEASTERN LOUISIANA UNIVERSITY

**WEST PUBLISHING COMPANY**

MINNEAPOLIS/ST. PAUL   NEW YORK   SAN FRANCISCO   LOS ANGELES

**WEST'S COMMITMENT TO THE ENVIRONMENT**

In 1906, West Publishing Company began recycling materials left over from the production of books. This began a tradition of efficient and responsible use of resources. Today, up to 95% of our legal books and 70% of our college texts and school texts are printed on recycled, acid-free stock. West also recycles nearly 22 million pounds of scrap paper annually—the equivalent of 181,717 trees. Since the 1960s, West has devised ways to capture and recycle waste inks, solvents, oils, and vapors created in the printing process. We also recycle plastics of all kinds, wood, glass, corrugated cardboard, and batteries, and have eliminated the use of Styrofoam book packaging. We at West are proud of the longevity and the scope of our commitment to the environment.

Production, Prepress, Printing and Binding by West Publishing Company.

 TEXT IS PRINTED ON 10% POST CONSUMER RECYCLED PAPER

# CONTENTS

**CHAPTER 1** HOMEOSTASIS: THE FOUNDATION OF PHYSIOLOGY 1

**CHAPTER 2** CELLULAR PHYSIOLOGY 9

**CHAPTER 3** MEMBRANE AND NEURONAL PHYSIOLOGY 21

**CHAPTER 4** CENTRAL NERVOUS SYSTEM 41

**CHAPTER 5** PERIPHERAL NERVOUS SYSTEM 59

**CHAPTER 6** MUSCLE PHYSIOLOGY 81

**CHAPTER 7** CARDIAC PHYSIOLOGY 97

**CHAPTER 8** BLOOD VESSELS AND BLOOD PRESSURE 113

**CHAPTER 9** BLOOD AND BODY DEFENSES 127

**CHAPTER 10** RESPIRATORY SYSTEM 151

**CHAPTER 11** URINARY SYSTEM 169

**CHAPTER 12** FLUID AND ACID-BASE BALANCE 183

**CHAPTER 13** DIGESTIVE SYSTEM 195

**CHAPTER 14** ENERGY BALANCE AND TEMPERATURE REGULATION 215

**CHAPTER 15** ENDOCRINE SYSTEM 223

**CHAPTER 16** REPRODUCTIVE SYSTEM 249

**APPENDIX** ANSWERS A-1

# HOMEOSTASIS:
# THE FOUNDATION OF PHYSIOLOGY

1

## CHAPTER OVERVIEW

The cells, the fundamental units of life, are organized into tissues, organs, and systems which in turn maintain homeostasis. This homeostasis is essential for the survival of the cells. The multicellular organism provides this discrete internal environment through its structural organization. The functions of tissues organs and systems are specializations of specific cellular capabilities. A cyclic phenomenon exists: The organism, its tissues, organs, and systems, is composed of cells maintained through the homeostasis of an internal environment which is controlled by body systems. This is the basic fundamental of physiology. Disruption of homeostasis can result in disease or even death.

## CHAPTER OUTLINE

### Level of Organization in the Body  (Text Page 2)

Cells are the basic unit of life (page 2).

The smallest unit capable of performing the basic functions essential for life is the cell. Whether as a unicellular organism or as part of a multicellular organism cells carry out these basic life processes. As part of a multicellular organism the cell has a dual role: Survival of the cell and survival of the organism. At the multicellular organismal level cells usually become specialized in one or several of the basic functions.

Cells are progressively organized into tissues, organs, systems, and, finally, the whole body (page 3).

Cells of similar origin, structure, and function are referred to as tissues. The four major types are muscles, nervous, epithelial and connective tissues.
There are three types of muscle tissue: Skeletal, cardiac and smooth muscle.

1

In nervous tissue, the brain, spinal cord, and various nerves of the body, the specialization is to convert information into electrical impulses, transmit these impulses to the
brain and subsequently to muscle or glands for the appropriate response.

Epithelial tissue lines or covers the body and its various organs, vessels, and cavities.
The specialization of this tissue is the exchange of materials. Highly specialized
epithelial cells form glands which provide essential secretions. Glands may be either
exocrine or endocrine.

Connective tissue, as its name implies, connects, supports and anchors the various structures of the body.

Organs are composed of several tissues organized to perform some specific function
within the body.

Those organs acting together to perform a basic life process comprise a system.

All the systems along with their organs tissues and cells make up the independent living
individual.

## Concept of Homeostasis (Text page 5)

Body cells are in contact with a privately maintained internal environment instead
of with the external environment that surrounds the body (page 5).

While each cell can perform most of the essential life processes, these basic functions
are more easily accomplished in a constant environment.

Since the external environment is continually changing , the cells of the body provide
survival skills to the whole organism through the various levels of organization which
result in a constant internal environment.

Body systems make exchanges with the external and internal environments as needed
by the cells.

This highly regulated internal environment is better suited to accommodating the needs
of the cells than the less stable external environment.

Homeostasis is essential for cell survival, and each cell, as part of an organized
system, contributes to homeostasis (page 5).

Body systems are comprised of cells progressively organized into tissues and organs.

Body systems perform the necessary functions to maintain homeostasis.

Homeostasis is essential for the survival of cells.

Homeostasis is a dynamic steady state.

There are eleven major body systems: Circulatory, digestive, respiratory, urinary,
skeletal, muscular, integumentary, immune, nervous, endocrine, and reproductive.

These eleven systems contribute to homeostasis by maintaining the following factors
in the internal environment: (1) Concentration of nutrients, oxygen, carbon dioxide,
waste products, water and electrolytes and (2) maintaining the pH, temperature,
pressure and volume.

Negative feedback is a common regulatory mechanism for maintaining
homeostasis (page 7).

Even the slightest changes in the internal environment must be detected and
controlled. There are two control systems for maintaining homeostasis: Intrinsic and
extrinsic. Intrinsic controls are built in or inherent to an organ. Extrinsic controls
utilize the nervous and endocrine systems.

## KEY TERMS

Anatomy-p2
Body systems-p4
Cell-p2
Connective tissue-p4
Endocrine glands-p4
Epithelial tissue-p3
External environment-p5
Extracellular fluid-p5
Extrinsic controls-p9

Exocrine glands-p4
Feedforward-p11
Glands-p4
Homeostasis-p5
Internal environment-p5
Interstitial fluids-p5
Intrinsic controls-p8
Lumen-p4
Multicellular organisms-p2

Muscle tissue-p3
Negative feedback-p10
Nervous tissue-p3
Organs-p4
Physiology-p2
Plasma-p5
Positive feedback-p11
Tissues-p3

## REVIEW EXERCISES

**True - False** (Answers on page A-1)

_____ 1. Tissues perform functions that are specializations of the normal function of individual cells.

_____ 2. Homeostasis maintains body systems to enhance the survival of cells.

_____ 3. Each body system is composed of different organs.

_____ 4. The endocrine system is a major control system.

_____ 5. The nervous system is an example of intrinsic control.

_____ 6. Feedforward mechanisms are frequently used in the digestive system.

_____ 7. All definitive actions initiated by nervous tissue are either muscle contractions or glandular secretions.

_____ 8. Cells require homeostasis, body systems maintain homeostasis, and cells make up body systems.

_____ 9. With positive feedback, a control systems input and output continue to enhance each other.

_____ 10. To sustain life, the internal environment must be maintained in an absolutely unchanging state.

_____ 11. Most human cells can be cultured; that is, when removed from the body, they will continue to thrive and reproduce in laboratory dishes when supplied with appropriate nutrients and other supportive materials.

_____ 12. Endocrine glands have ducts.

_____ 13. Control systems that operate to maintain homeostasis can be grouped into two classes -- intracellular and extracellular.

_____ 14. Humans are multicellular organisms.

_____ 15. Connective tissue is distinguished by having relatively few cells dispersed within an abundance of extracellular material.

**Fill in the Blank** (Answers on page A-1)

16. _____ glands have ducts.
17. A _____ is the cavity within the interior of hollow organs and tubes.
18. _____ is essential for the survival of cells in the body.
19. _____ fluid is part of the extracellular fluid.
20. The _____ system transfers water and electrolytes from the external environment into the internal environment.
21. _____ is a common regulatory mechanism for maintaining homeostasis.
22. The smallest unit capable of carrying out the processes associated with life is the _____.
23. The body cells are in direct contact with and make life-sustaining exchanges with the _____.
24. The internal environment consists of the _____, which is made up of _____, the fluid portion of the blood, and _____ _____ which surrounds and bathes all cells.
25. _____ exists when a change in a regulated variable triggers a response that opposes the change.
26. _____ refers to the abnormal functioning of the body associated with disease.
27. The basic unit of both structure and function is the _____.
28. _____ is a naturally occurring protein that promotes development of nervous tissue and skin in embryos.
29. _____ are epithelial-tissue derivatives that are specialized for secretion.
30. Organs are further organized into _____, each of which is a collection of organs that perform related functions and interact to accomplish a common activity that is essential for survival of the whole body.
31. There are eleven major body systems. List them (1) _____,
    (2) _____, (3) _____,
    (4) _____, (5) _____,
    (6) _____, (7) _____,
    (8) _____, (9) _____,
    (10) _____, (11) _____.

**Matching**: Match organ to body system. (Answers on page A-2)

_____ 32. Skin
_____ 33. Joints
_____ 34. Trachea
_____ 35. Adrenals
_____ 36. Special sense organs
_____ 37. Uterus
_____ 38. Tonsils
_____ 39. Skeletal muscles
_____ 40. Blood
_____ 41. Urethra
_____ 42. Liver

a. Reproductive
b. Circulatory system
c. Digestive system
d. Immune system
e. Integumentary system
f. Muscular system
g. Skeletal system
h. Endocrine system
i. Urinary system
j. Nervous system
k. Respiratory system

**Multiple Choice** (Answers on page A-2)

_____ 43. These cells, when destroyed by trauma or disease, cannot be replaced.
   a. glandular tissue
   b. epithelial tissue
   c. muscles and nerves
   d. connective tissue
   e. none of the above

_____ 44. Which system is important in maintaining the proper pH of the internal environment?
   a. circulatory
   b. digestive
   c. muscular
   d. respiratory
   e. immune

_____ 45. This tissue lines the stomach.
   a. connective
   b. muscle
   c. epithelial
   d. nervous
   e. mesodermal

_____ 46. Which type of control is exemplified by an air conditioner, its thermostatic device, and their electrical connections?
   a. positive feedback
   b. negative feedback
   c. feedforward
   d. biofeedback
   e. cybernetic

_____ 47. Which muscle type moves the contents of food through the intestines?
- a. cardiac
- b. skeletal
- c. smooth
- d. extrinsic
- e. intrinsic

_____ 48. Which is not a basic cell function?
- a. Maintain a storage reservoir for potassium which will maintain a relatively constant concentration of potassium in the extracellular fluid.
- b. Perform various chemical reactions that use nutrients and oxygen to provide energy for the cell.
- c. Obtain food and oxygen from the environment surrounding the cell.
- d. Be sensitive and responsive to changes in the environment surrounding the cell.
- e. Control to a large extent the exchange of materials between the cell and its surrounding environment.

_____ 49. Which type of tissue utilizes the inherent capability of cells to produce intracellular movement?
- a. epithelial
- b. muscle
- c. nervous
- d. connective
- e. endoderm

_____ 50. Which type of tissue utilizes the inherent capability to synthesize digestive enzymes?
- a. epithelial
- b. muscle
- c. nervous
- d. connective
- e. glandular

_____ 51. During the minute that it will take to answer this question, all but one of these activities will be performed.
- a. your body will use about 2 calories of energy
- b. your body will exchange 6 liters of air between the atmosphere and your lungs
- c. more than 1 liter of blood will flow through your kidneys
- d. your heart will beat 70 times, pumping 5 liters of blood to your lungs and another 5 liters to the rest of your body
- e. your brain will process 325 afferent impulses

_____ 52. Brain cells were first cultured in a laboratory in
      a. 1990
      b. 1940
      c. 1890
      d. 1965
      e. 1955

_____ 53. Factors of an internal environment that must be homeostatically maintained include
      a. its pH and temperature
      b. its concentration of water, salt, and other electrolytes
      c. its volume and pressure
      d. all of the above
      e. none of the above

_____ 54. Which of the following is a component of the immune system?
      a. appendix
      b. adrenals
      c. bulbourethral glands
      d. hypothalamus
      e. none of the above

_____ 55. Which system controls and coordinates bodily activities that require swift responses?
      a. skeletal system
      b. muscular system
      c. reproductive system
      d. nervous system
      e. endocrine system

_____ 56. What feedback continually enhances the output so that the controlled variable continues to be moved in the direction of the initial change?
      a. negative
      b. positive
      c. feedforward
      d. biofeedback
      e. cybernetic

_____ 57. What controls are built in or inherent to an organ?
      a. intrinsic
      b. extrinsic
      c. endocrine
      d. exocrine
      e. holocrine

_____ 58. What controls are regulatory mechanisms  initiated  outside of an organ to alter the activity of the organ?
    a. intrinsic
    b. extrinsic
    c. mesocrine
    d. epicrine
    e. holocrine

_____ 59. What controls use the nervous and endocrine systems to control various organs and systems?
    a. intrinsic
    b. extrinsic
    c. mesocrine
    d. holocrine
    e. exocrine

**Points to Ponder**
1. Why does the heart rate of a well trained athlete begin to increase a few moments before the competitive event occurs?
2. As you sit in class you start yawning.  Looking around, you see other students doing the same.  What do you think is happening physiologically?
3. How would you rate homeostasis as a diagnostic tool?
4. After studying this chapter on homeostasis why do you think it is important for people to have thorough examinations by their physicians?

# CELLULAR PHYSIOLOGY    2

## CHAPTER OVERVIEW

The basic processes associated with life require cellular functions at the
molecular level. These molecular activities are performed by various
organelles of the cell. Some organelles such as mitochondria and
lysosomes act directly in the basic life processes while cytoskeletal
structures function in areas of structural integrity, maintenance, and movement. The
molecular activity of the organelles and other cellular components is essential for the
proper functioning of tissues, organs and body systems in the maintenance of
homeostasis. Energy produced in organelles drives all the body systems.
The role of a body system in homeostasis is inseparable from the basic
molecular functions of its cells.

## CHAPTER OUTLINE

### Introduction (Text page 15)

Cells are the bridge between molecules and humans (page 15).

Even though the chemicals comprising a cell have been thoroughly analyzed
researchers have not yet organized these chemicals into a cell in the laboratory.
Cells are the building blocks for whole body.

Increasingly better tools are revealing the complexity of cells (page 15).

Human cells average about 10 to 20 micrometers in diameter.
Lined up side by side, about one hundred average-sized cells would occupy a
distance of only 1 millimeter.

Cells were unknown until the middle of the seventeenth century.
The electron microscope has revealed the great diversity and complexity of the
 internal structure of the cell.
Through the availability of better microscopy, biochemical techniques, cell-culture
technology, and genetic engineering, the cell has emerged  as a complex highly
organized, compartmentalized structure.

## Most cells are subdivided into plasma membrane, nucleus and cytoplasm (page15).

The thin membranous structure which surrounds the cell is the plasma membrane or
 cell membrane which serves as the boundary of the cell.
The fluid within the cell is the intracellular fluid (ICF) while fluids outside the cell are
referred to as extracellular fluid (ECF).
The plasma membrane selectively controls the movement of molecules between the ICF
and ECF.
The plasma membrane also recognizes other cells and foreign materials.
The nucleus, surrounded by the double-layered nuclear membrane, is the largest single
organized cellular component.
Deoxyribonucleic acid (DNA) is found in the nucleus.
DNA controls  protein synthesis.
Since most cellular activities involve structural or enzymatic proteins, DNA provides
the instructions and thus the control of these activities.
As a genetic blueprint DNA ensures cellular reproduction and organismal reproduction.
That part of the cell outside the nucleus is called the cytoplasm and is composed of a
gel-like mass known as cytosol.
There are six organelles dispersed in the cytosol:
Endoplasmic reticulum, golgi complex, lysosomes, peroxisomes, mitochondria and
vaults.
The cytosol is laced with an elaborate protein network which constitutes the cyto-
skeleton.
The cytoskeleton  provides  shape  to the cell, internal organization, and internal
movement.

# Organelles (Text page 17)

## The endoplasmic reticulum is a synthesizing factory (page 17).

There are two types of endoplasmic reticulum (ER); smooth and rough.
The ER is one continuous membranous organelle with many interconnected channels.
The rough ER derives its name from dark-staining particles, ribosomes, which are
attached to the ER membrane.
Ribosomes are composed of ribonucleic acid (RNA) protein complexes.
The rough ER and its associated ribosome synthesize proteins and release these products
into the ER lumen.
Some of these new proteins are destined for export to the exterior of the cells as a
secretion.
Other new proteins are used within the cell as structural proteins or enzymes.
Cellular membranes consist predominantly of lipids and proteins.
The lipids necessary for the production of new membranes are synthesized using enzymes
contained in the membranous wall of the ER.
The new proteins and lipids are passed into the ER lumen.

Molecules destined for export out of the cell are separated in this manner.
Having no ribosomes smooth ER is not involved in protein synthesis.
Newly synthesized proteins are transported to the smooth ER where they
"bud off" and become membrane enclosed transport vesicles.
These transport vesicles move to the Golgi complex for further processing.

## The Golgi complex is a refining plant and directs molecular traffic (page 20).

The Golgi complex consists of membranous, flattened, slightly curved sacs stacked in
layers called Golgi stacks.
Transport vesicles enter the Golgi stacks at a point nearest the center of the cell.
The Golgi complex perform two important interrelated functions: Processing the raw
materials into finished products and sorting and directing the finished products to
their final destination.
Some vesicles are destined for intracellular use.
Secretory vesicles are destined to be released to the exterior of the cell in a process called
exoctyosis.
All vesicles bud off the dilated edges of Golgi sacs.

## Lysosomes serve as the intracellular digestive system (page 21).

Lysosomes are membrane-bound sacs containing powerful hydrolytic enzymes.
Extracellular material is bound to the plasma membrane which then invaginates forming
a membrane-bound intracellular vesicle through a process known as Endocytosis.
White blood cells use a special form of endocytoses called phagocytosis to trap
multi-molecular particles.
Lysosomes fuse to the intracellular vesicle and release their hydrolytic enzymes into the
vesicle.
Lysosomes become residual bodies upon completion of their digestive activities and are
eliminated by exocytosis.
Storage diseases result if an enzyme is missing in the lysosome.
Abnormal accumulation of gangliosides produces the condition known as Tay - Sachs
Disease.

## Peroxisomes house oxidative enzymes that detoxify various wastes (page 22).

Peroxisomes are membrane-bound sacs containing oxidative enzymes that strip hydrogen
from toxic wastes.
Ethanol, consumed in alcoholic beverages is detoxified in this manner.

## Mitochondria are energy organelles (page 22).

Mitochondria are rod or oval-shaped organelles.
Mitochondria have smooth outer membranes and inner membranes that have
infoldings, or cristae, into the gel-like matrix.
Cristae are the site of the electron transport chain activity.
The citric acid cycle occurs in the matrix.

## Energy derived from food is stored in ATP (page 23).

Usable energy for the cell is generated by splitting the terminal phosphate from
adenosine triphosphate (ATP) thereby producing adenosine diphosphate (ADP).
Glucose from the digestive system via the blood is transported across the plasma

membrane into the cytosol where it is converted to pyruvic acid and two molecules of ATP through the chemical process glycolysis.

Pyruvic acid is converted to acetyl coenzyme A in the matrix of the mitochondria.

Acetyl coenzyme A joins oxaloacetic acid in the citric acid cycle which produces two molecules of carbon dioxide, oxaloacetic acid and one molecule of ATP.

The citric acid cycle is also known as the Krebs cycle.

Hydrogen is removed using carrier molecules nicotinamide adenine dinucleotide (NAD) and flavine adenine dinucleotide (FAD).

One molecule of glucose produces two acetic-acid molecules and two turns of the citric acid cycle yielding two molecules of ATP.

High energy electrons are extracted from hydrogen bonds with NAD and FAD producing thirty-two molecules of ATP.

At three sites in the electron transport chain ATP is produced using oxygen in a process known as oxidative phosphorylation.

Glycolysis is anaerobic.

Since the citric acid cycle and the electron transport chain are aerobic, and oxygen insufficiency will not impede glycolysis, excess pyruvic acid is converted to lactic acid through a reversible reaction.

## The energy stored within ATP is used for synthesis, transport, and mechanical work (page 26).

Cellular activities that require energy expenditure fall into three main categories:

(1) Synthesis of new compounds such as protein synthesis by the ER.

(2) Membrane transport such as selective transport in the kidney.

(3) Mechanical work such as muscular contraction.

## Vaults are a newly discovered organelle (page 29).

Vaults are three  times as large as ribosomes and are shaped like octagonal barrels.

These organelles may be used in transport of messenger RNA.

Since vaults are abundant in areas where actin is assembled they are suspected to be somehow involved in cellular contractile systems.

# Cytosol and cytoskeleton (Text page 29)

## The cytosol is important in intermediary metabolism, ribosomal protein synthesis, and storage of fat and glycogen (page 30).

Intermediary metabolism refers to the large set of intracellular chemical reactions that involve degradation, synthesis and transformation of small organic molecules such as simple sugars, amino acids and fatty acids.

Glycolysis is such a reaction.

Throughout the cytosol, free ribosomes are dispersed that synthesize proteins for use in the cytosol.

Glycogen and fat are stored in the cytosol as inclusions.

The cytoskeleton supports and organizes the intracellular components and controls their movements (page 30).

The cytoskeleton is found in the cytosol and is composed of four distinct elements: Microtubules, microfilaments, intermediate filaments and the microtrabecular lattice.

Microtubules are long slender tubes composed of the small globular proteins, tubulin.

Microtubules are necessary for maintaining asymmetrical cell shapes such as in the axon of a neuron which may be up to a meter in length.

Microtubules are essential in the transport of secretory vesicles from one region of a cell to another.

The distribution of chromosomes during cellular division depends on the formation of the mitotic spindle by microtubules.

Microtubules are the dominate structure and functional components of cilia and flagella.  Cilia are numerous tiny protrusions.  Flagella are long whiplike structures.

Humans have cilia on the cells lining the trachea, with cilia also lining the oviducts in females.

Sperm tails are actually flagella.

Microfilaments are composed of actin in most cells and also myosin in muscle cells.

Actin microfilaments are two twisted strings of the globular protein actin.

Actin microfilaments are used to produce amoeboid movement necessary for locomotion of fibroblasts, skin cells and white blood cells.

Microfilaments serve as mechanical stiffeners in the microvilla which line the free edge of epithelial cells lining the intestine and kidney tubules.

These microvilla increase the absorptive surface area.

The hair cells of the inner ear are specialized microvilla.

Intermediate filaments derive their name by being between microtubules and microfilaments in size.

Keratin-containing intermediate filaments form irregular networks which connect to extracellular filaments to strengthen and waterproof the skin.

The microtrabecular lattice provides a dynamic structural framework within the cytoplasm.

This lattice of exceedingly fine filaments serves to suspend organelles, microtubules, and free ribosomes.

These filaments extend from the inner layer of the plasma throughout the cytoplasm.

# KEY TERMS

Actin-p31
Adenosine diphosphate-p23
Adenosine triphosphate-p23
ATP synthetase-p26
Citric-acid cycle-p23
Deoxyribonucleic acid-p15
Electron transport chain-p26
Endocytosis-p21
Endoplasmic reticulum-p17

Flavin adenine dinucleotide-p26
Golgi complex-p20
Lysosomes-p21
Microfilament-p31
Microtrabecular lattice-p33
Microtubules-p31
Myosin-p31
Nicotinamide adenine
    dinucleotide-p26

Oxidative phosphorylation-p26
Phagocytosis-p21
Ribonucleic acid-p15
Ribosomes-p17
Secretary vesicles-p21
Transport vesicles-p19
Tubulin-p31

## REVIEW EXERCISES

**True - False** (Answers on page A-3)

_____  1. Tay-Sachs in a disease in which brain cells cannot perform glycolysis.

_____  2. The electron transport chain in the mitochondrial matrix produces thirty-two molecules of ATP per molecule of glucose processed.

_____  3. Ketoglutaric acid is part of the citric acid cycle.

_____  4. Glycolysis is a good example of oxidative phosphorylation.

_____  5. The chemiosmotic mechanism is another name for the tricarboxylic acid cycle.

_____  6. Carbon dioxide produced by the Krebs cycle is released into the atmosphere.

_____  7. Rough ER with its ribosomes synthesize proteins.

_____  8. "Life" is due to the complex organization and interaction of chemical molecules within the cells.

_____  9. The cells of the human body are surrounded by a cell wall.

_____ 10. All body functions ultimately depend on the activities of the individual cells that compose the body.

_____ 11. About 100 average-sized cells lined up side by side would stretch a distance of only 1 mm.

_____ 12. The plasma membrane does not merely serve as a mechanical barrier to hold in the contents of the cell; it has the ability to selectively control movement of molecules between the ICF and ECF.

_____ 13. The nucleus is the largest multicellular component usually located near the center of the cell.

_____ 14. Within the nucleus is the cell's genetic material, RNA.

_____ 15. Organelles are like intracellular "specialty shops".

_____ 16. Protein synthesis is carried out by lysosomes.

_____ 17. All ribosomes are produced in the cell's nucleus under the direction of DNA, with each being "programmed" at any given time to facilitate the synthesis of only one specific protein needed by that particular cell.

_____ 18. The smooth ER contains ribosomes.

_____ 19. In liver cells, the smooth ER has a special capability.  It contains enzymes that are involved in detoxifying harmful substances produced within the body by metabolism or substances that enter the body from the outside in the form of drugs or other foreign compounds.

_____ 20. The Golgi complex consist of sets of  flattened, slightly curved, membrane-bound sacs, or cisternae, stacked in layers.

_____ 21. Coated vesicles originate from the dilated rims of the nucleus.

_____ 22. Release of the contents of the secretory vesicle constitutes the process of secretion.

_____ 23. Residual bodies are lysosomes that have completed their digestive activities.

_____ 24. Each mitochondrion is enclosed by a single membrane.

_____ 25. Energy derived from food is stored in the mitochondria.

_____ 26. In glycolysis, 1 molecule of glucose has a net yield of only 32 molecules of ATP.

_____ 27. The newly identified sixth type of organelles are called vaults.

_____ 28. The cytosol is the semiliquid portion of the cytoplasm that surrounds the organelles.

_____ 29. The endoplasmic reticulum is one continuos organelle consisting of many tubules and cisternae.

_____ 30. Intermediate filaments play a structural role in parts of the cell subject to mechanical stress.

_____ 31. Microfilaments are found in microvilli.

_____ 32. Intermediate filaments are a functional component of cilia.

_____ 33. Transport vesicles release their contents in a process known as secretion.

_____ 34. Peroxisome contain reductive enzymes which strip hydrogen from specific molecules.

_____ 35. Glycolysis occurs in the mitochondria.

_____ 36. The citric acid cycle, Krebs cycle and tricarboxylic acid cycle are all the same.

_____ 37. Pyruvic acid is one of the acids of the citric acid cycle.

**Fill in the Blank** (Answers on page A-4)

38. The _____ RNA exits the nucleus through nuclear pores.

39. The _____ "bud off" from the smooth ER.

40. The _____ appears to suspend microtubules and microfilaments, as well as various organelles.

41. The _____ RNA "reads" the code and translate it into the appropriate amino acid sequence.

42. _____ . are membranous sacs containing hydrolytic enzymes.

43. The _____ is a specialized cytoplasmic structure that forms cilia.

44. Oxidative enzymes are found in the _____.

45. A cell is made up of three major subdivisions: (1)_____,
(2) _____, and (3) _____.

46. The cytosol is pervaded by a protein scaffolding called _____., that serves as the "bone and muscle" of the cell.

47. The ribosomes of the rough ER synthesize _____, whereas its membranous walls contain enzymes essential for the synthesis of _____.

48. The _____ are shaped like octagonal barrels.

49. The _____ ER is the central packaging and discharge site for molecules to be transported from the ER.

50. The process in which a secretory vesicle fuses with the plasma membrane, then opens up and extrudes its contents to the exterior is known as _____.

51. Lysosomes contain (what type of) _____ enzymes.

52. Foreign material to be attacked by lysosomal enzymes is brought into the cell by the process of _____ .
53. _____ are helically intertwined chains of actin molecules.
54. The nucleus of the cell contains _____, the cells genetic material.
55. _____ help maintain asymmetrical cell shapes.
56. The _____ is a very thin membranous structure that encloses each cell.

**Matching**: Match component or reaction to the organelle or structure in which it occurs.
        (Answers on page A-4)

| | |
|---|---|
| _____ 57. Electron transport chain | a. cytosol |
| _____ 58. Oxidative enzymes | b. golgi complex |
| _____ 59. Microtubules | c. microvilli |
| _____ 60. Glycolysis | d. mitochondrial matrix |
| _____ 61. Hydrolytic enzymes | e. ribosome |
| _____ 62. Transport vesicles | f. nucleus |
| _____ 63. Secretory vesicles | g. mitochondrial inner membrane |
| _____ 64. Microfilaments | h. smooth ER |
| _____ 65. DNA | i. lysosomes |
| _____ 66. RNA | j. flagella |
| _____ 67. Citric acid cycle | k. peroxisomes |

**Multiple Choice** (Answers on page A-4)
68. In mitosis the last phase involves the constriction of the cytoplasm to produce two daughter cells. Which of the following is involved in this cytoplasmic constriction?
        a. clathrin
        b. actin
        c. kinesin
        d. malic acid
        e. dynein
69. Which of the following is NOT part of the citric acid cycle?
        a. citric acid
        b. succinic acid
        c. fumaric acid
        d. malic acid
        e. pyruvic acid

70. Tay-Sachs disease is a storage disease involving accumulations of these compounds.
    a. gangliosides
    b. oxidative enzymes
    c. hydrogen peroxide
    d. plasma gel
    e. ribophorins

71. Which is the correct order for the citric-acid cycle?
    a. oxaloacetic acid--citric acid--isocitric acid--ketoglutaric acid--fumaric acid--succinyl CoA--succinic acid--malic acid
    b. oxaloacetic acid--citric acid--isocitric acid--ketoglutaric acid--succinyl CoA--succinic acid--fumaric acid--malic acid
    c. citric acid--isocitric acid--oxaloacetic acid--ketoglutaric acid--succinyl acid--succinic acid--malic acid--fumaric acid
    d. oxaloacetic acid--isocitric acid--citric acid--succinyl CoA--succinic acid-ketoglutaric acid--fumaric acid--malic acid
    e. oxaloacetic acid--citric acid--isocitric acid--ketoglutaric acid--succinic acid--succinyl CoA--fumaric acid--malic acid

72. Which compound is used as mechanical stiffeners in microvilli?
    a. actin
    b. kinesin
    c. dynein
    d. myosin
    e. clathrin

73. Which is composed of RNA and proteins?
    a. chromosomes
    b. peroxisomes
    c. lysosomes
    d. ribosomes
    e. cortisone

74. Which of the following functions as the modifier, packager and distributor for newly synthesized proteins?
    a. vaults
    b. Golgi complex
    c. secretory vesicles
    d. endoplasmic reticulum
    e. mitochondria

75. Which are described as octagonal barriers?
    a. vaults
    b. Golgi complex
    c. residual bodies
    d. free ribosomes
    e. mitochondria
76. Which components are specialized to detect sound?
    a. microtubules
    b. intermediate filaments
    c. microtrabecular lattice
    d. spindle fibers
    e. microfilaments
77. Tubulin is found in which of the following?
    a. cilia and flagella
    b. cilia and microvilli
    c. flagella and brush borders
    d. mitotic spindles and microtrabecular lattices
    e. none of the above
78. Which of the following processes acetyl CoA?
    a. electron transport chain
    b. citric acid cycle
    c. glycolysis
    d. respiratory chain
    e. fermentation
79. Which of the following converts glucose into two pyruvic acid molecules?
    a. electron transport chain
    b. glycolysis
    c. citric acid cycle
    d. respiratory chain
    e. transition reaction
80. Which cellular structure organizes the glycolytic enzymes in sequential alignment?
    a. microtubules
    b. microfilaments
    c. intermediate filaments
    d. microtrabecular lattice
    e. vaults

81. Which of the following are the smallest elements visible with a conventional electron microscope?
    a. microtubules
    b. microfilaments
    c. intermediate filaments
    d. microtrabecular lattice
    e. vaults

82. Which ribosome synthesizes proteins used to construct new cell membranes?
    a. free ribosome
    b. rough ER bound ribosome
    c. smooth ER
    d. all of the above
    e. none of the above

83. Which ribosome synthesizes proteins used intracellularly within the cytosol?
    a. free ribosome
    b. rough ER bound ribosome
    c. neither
    d. both

84. Which of the following takes place in the mitochondrial matrix?
    a. glycolysis
    b. citric acid cycle
    c. electron transport chain
    d. fermentation
    e. cytolysis

85. Which reaction produces water as a by-product?
    a. electron transport chain
    b. citric acid cycle
    c. glycolysis
    d. calvin reaction
    e. fermentation

86. Which of the following is **not** characteristic of the cytoskeleton?
    a. supports the plasma membrane and is responsible for the particular shape, rigidity, and spatial geometry of each different cell type
    b. plays a role in regulating cell growth and division
    c. elements are all rigid, permanent structures.
    d. is responsible for cell contraction and cell movement
    e. supports and organizes the ribosomes, mitochondria, and lysosomes

87. This structure has hairlike motile protrusions
    a. flagella
    b. cilia
    c. actin bundles
    d. sperm
    e. fermentation

## Points to Ponder
1. What conditions would prompt primitive bacteria or early mitochondria to invade primitive cells and become mitochondria as we know them?
2. Aside from energy production what is the advantage in having the citric acid cycle?
3. How do you think your cells replace nicotinamide adenine dinucleotide?
4. Would it be possible to calculate the ATP production knowing how much carbon dioxide was produced?

## Clinical Perspectives
1. In carbon monoxide poisoning, the gas carbon monoxide readily combines with hemoglobin. How does this phenomenon produce such fatigue that the victim may not be able to escape the gas?

# MEMBRANE AND

# NEURONAL PHYSIOLOGY

## CHAPTER OVERVIEW

The plasma membrane is a lipid bilayer with membrane proteins, membrane carbohydrates and cholesterol arranged as a fluid mosaic model. Some proteins form channels across the lipid bilayer, others serve as carrier molecules and still others are receptor sites on the outer surface of the cell. Using these proteins and several enzymes molecules can enter the cell. Cells can be held together using fibrous proteins: (1) collagen, (2) elastin and (3) fibronectin. Cells can be directly linked together using (1) desmosomes, (2) tight junctions and (3) gap junctions. Two forces are used to transport molecules across a membrane (1) passive and (2) active. Diffusion is passive down a concentration gradient. Ions may use an electrical gradient for transport. Osmosis is water diffusing down its own concentration gradient. Larger molecules use carrier-mediated transport. The sodium-potassium pump is such a mechanism that performs several functions for the cell. Vesicular transport is a special situation to gain, secrete, rebuild or destroy various entities. The membrane potential and the various transport mechanisms demonstrate the importance of the plasma membrane in maintaining homeostasis. As changes occur in the external and internal environments of the body, the nerve cells receive this information and transmit it to other cells. Eventually nerve cells initiate the necessary responds to such changes. Therefore, neurons are essential in maintaining homeostasis.

The process of transmission of information within a neuron involves electrical signals such as grades and action potentials. Through changes in permeability the nerve cell undergoes depolarization, hyperpolarization and repolarization. These phenomena trans-mit the information. The speed of transmission depends on the diameter of the fiber and whether or not the neuron is myelinated. When the action potential reaches the synaptic knob a neurotransmitter is released which diffuses across the synaptic cleft and changes the permeability of the postsynaptic neuron.  Through temporal or spatial summation an action potential is initiated in the axon hillock of the postsynaptic neuron.  Synapses may be excitatory or inhibitory.  Drugs and diseases may alter synaptic effectiveness.

# CHAPTER OUTLINE

## Membrane Structure and Composition (Text Page 38)

The plasma membrane separates the intracellular and extracellular fluid (page 38).

Each cell type is unique in its intracellular contents.

The plasma membrane is a thin layer of lipids and proteins forming the outer boundary of the cell.

This membrane maintains the specific composition of the intracellular fluid by selectively permitting various substances to pass between the cell and the extracellular fluid.

The plasma membrane is a fluid lipid bilayer embedded with proteins (page 38).

All plasma membranes are composed of lipids, proteins and carbohydrates.

The lipids are predominately phospholipid with some cholesterol.

The negatively charged polar head of the phospholipid contains a phosphate group.

The polar head is hydrophilic.

The two fatty acids make up the polar tails and are not electrically charged.

The tails are hydrophobic.

The two lipid layers are arranged such that the outer polar heads are in contact with the extracellular fluid and the inner polar heads are in contact with the intracellular fluid.

The hydrophobic tails of each lipid layer are toward the inside of the plasma membrane.

The fluid nature of the membrane is maintained by the lack of bonding between phospholipids and cholesterol between the fatty acids.

Membrane proteins may extend through the thickness of the membrane while others are found only on the inner and outer surfaces.

Short-chain carbohydrates and membrane carbohydrates are bound to the outer membrane proteins and lipids.

The molecular arrangements of plasma membrane is called the fluid mosaic model.

The lipid bilayer forms the primary barrier to diffusion, whereas proteins perform most of the specific membrane functions (page 39).

The lipid bilayer provides three important functions:  (1) It forms a structural barrier around the cell.  (2) It prevents passage of water-soluble substances between the ICF and ECF.  (3) It provides fluidity to the membrane.

Membrane proteins provide seven specialized functions:  (1) Proteins spanning the membrane form highly selective channels which allow or prevent passage of small particles such as ions.  (2) Other proteins serve as carrier molecules to transport substances between the ECF and ICF.  (3) Proteins on the outer surface may serve as a receptor sites for specific molecules in the ECF. (4) Proteins also function as membrane-bound enzymes to control surface chemical reactions. (5) Other proteins on the inner surface form a filamentous meshwork and attach to the cytoskeleton. (6) Additional proteins function on the outer surface as cell adhesion molecules (CAMs) and are used to attach cells and tissues together. (7) Still other proteins, functioning with special carbohydrates, serve to recognize other cells.

There are three possible functions for outer surface membrane carbohydrates: (1) They may serve to orient and anchor membrane proteins. (2) Surface membrane carbohydrates possibly function with membrane proteins to recognize cells. (3) Still other carbohydrates appear to be markers for tissue boundaries.

## Membrane Receptors and Postreceptor Events (Text page 42)

Binding of chemical messengers to membrane receptors brings about a range of responses in the different cells though only a few remarkably similar pathways are used (page 42).

As selected chemical messengers in the ECF come in contact with the cell they are bound to specialized protein receptors on the outer surface of the plasma membrane. There are two methods by which the desired intracellular responses are achieved using the binding of the protein receptor to the extracellular chemical messenger or first messenger: (1) By opening or closing channels through the membrane or (2) by transferring the signal to an intracellular chemical messenger or second messenger.

If the first messenger opens or closes the sodium and potassium channels, the electrical activity of cells that generate electrical signals is altered.

Another intracellular response occurs if the calcium channels are opened.

Cytosolic calcium increases, triggering the release of secretion in many gland cells. Another method is to alter sodium and potassium channels which will set up an electrical impulse within the cell and thereby open the calcium channels.

The release of chemicals from nerve cells in response to a nerve impulse is such an example.

Still another method involving the electrical impulse within the cell as just mentioned will open calcium channels within organelles.

This increase in cytosolic calcium sets in motion events leading to a muscle contraction.

There are two second messenger pathways.

The extracellular messenger binds to a protein receptor which causes the activation of an enzyme on the inner surface which in turn activates the intracellular messengers.

The intracellular messenger can be cyclic adenosine monophosphate (cyclic AMP) or calcium.

Using the cyclic AMP pathway, the extracellular messenger uses the membrane activates the enzyme adenylate cyclase which converts ATP to cyclic AMP by splitting off two phosphates.

Cyclic AMP triggers the intracellular event by activating cyclic AMP-dependent protein kinase which phosphorylates the specific enzyme.

The altered enzyme performs its function thus completing the dictated response.

Activation of the cyclic AMP system brings about a wide range of responses: (1) Modifies heart rate in the heart, (2) Stimulates sex hormones formation by the ovaries, and (3) Controls water conservation in the kidneys.

Many disease processes such as myasthenia gravis can be linked to malfunctioning receptors or defects in one of the components of the various pathways.

# Cell-to-Cell Adhesions (Text page 43)

## The extracellular matrix serves as the biological "glue" (page 44).

The plasma membrane functions in cell-to-cell adhesions that permit groups of cells to bind together into tissues and organized further into organs.
The extracellular matrix is a meshwork of fibrous proteins in a watery gel composed of complex carbohydrates.
The protein fibers are collagen, elastin and fibronectin.
The extracellular matrix is secreted by fibroblasts.

## Some cells are directly linked together by specialized cell junctions (page 44).

At a desmosome filaments extend between the plasma membranes of two closely adjacent but non-touching cells to anchor the cells together.
Sheets of epithelial tissue are joined by tight junctions.
Adjacent cells are joined together in a tight seal which is impermeable and thereby prevents materials from passing between cells.
Selected substances must pass through the cells.
Gap junctions allow particles such as ions to pass between cells whereas larger molecules are blocked.
Connexons are small connection tunnels between to adjacent cells attached together by a gap junction.

# Membrane Transport (Text page 46)

## Lipid-soluble substances and small ions can passively diffuse through the plasma membrane down their electrochemical gradient (page 46).

The membrane is permeable to the substance if the substance can cross the membrane and impermeable if the substance cannot cross.
Selectively permeable membranes permit some particles to pass but exclude others.
Two properties of particles determine whether they can cross the membrane without assistance. (1) Particles must be soluble in lipids. (2) Particles must be small.
Forces are required to move particles across the membrane.
Passive forces do not require the cell to expend energy to produce movement.
In conditions where the cell must expend energy the movement is achieved by active forces.
A difference in concentration between two adjacent areas is referred to a concentration gradient.
Diffusion uses passive forces to move particles down a concentration gradient.
While diffusion involves movement of molecules in all directions the net movement of a substance is from an area where the concentration of the substance is greater to an area of lesser concentration.
When the molecules of the substance are uniformly distributed, a state of equilibrium or steady state exists.
The difference between two opposing movements is referred to as new diffusion.
Fick's law of diffusion demonstrates the effects of several factors on rate of net diffusion. (1) Greater magnitude of the concentration gradient increases the rate of net diffusion. (2) Increased permeability of the membrane to the substance increases the

rate of net diffusion.  (3) The rate of net diffusion is also increased by an increased surface area of membrane.

The next two factors decrease the rate of net diffusion:  (1) increased molecular weight and (2) increased thickness of the membrane.

Diffusion also moves substances along an electrical gradient.

Positively charged ions (cations) move toward negatively charged areas whereas negatively charged ions (anions) move toward positively charged areas.

If an electrical gradient (difference in charge) exists between the ICF and ECF, and the membrane is permeable to the ions in question then these ions will diffuse along this electrical gradient.

The simultaneous existence of an electrical gradient and a concentration gradient for a specific ion is know as an electrochemical gradient.

## Osmosis is the net diffusion of water down its own concentration gradient (page 48).

Osmosis is a special type of diffusion involving a water-soluble solute and water.

If the solute cannot pass through the membrane, the water uses its own concentration gradient to dilute the solute on the other side.  In this type of osmosis the volume on the side of the solute increases.

If both solute and water can easily pass through the membrane then both will cross the membrane until each has the same concentration on both sides.

The resulting volume will be the same as at the onset of the diffusion.

As water continues to cross the membrane a hydrostatic pressure develops which increases as the volume increases.

Hydrostatic pressure opposes osmosis.

Osmotic pressure is equal to the hydrostatic pressure necessary to stop the osmosis.

## Special mechanisms are used to transport selected molecules unable to cross the plasma membrane on their own (page 50).

Carrier-mediated transport is used for large, poorly lipid-soluble molecules such as glucose, proteins and amino acids which cannot cross the plasma membrane.

This impermeability ensures that large essential molecules cannot escape from the cell.

The carrier proteins span the thickness of the plasma membrane.

These proteins change shape to expose the binding sites to each surface of the membrane alternately.

The substance to be transported attaches to a carrier protein that is specific for one substance or at most a few closely related compounds.

The transport maximum of a substance is the upper limit that can be transported per unit time.

This transport maximum occurs when all the carriers are being used,  thus saturation.

This saturation condition can be overcome by changing the affinity of the binding site for the passenger or by increasing the number of binding sites.

Closely related chemical compounds compete for binding sites.

Carrier-mediated transport takes two forms:  (1) Facilitated diffusion goes down a concentration gradient and is passive.  (2) Active transport goes against the concentration gradient.

In active transport the carrier is phosphorylated on the low concentration side to enhance affinity of binding site for passenger.

Carrier is dephosphorylated on high concentration side to reduce affinity.

These active transport mechanisms are called pumps.

The sodium-potassium pump is such a mechanism that performs two important roles:
(1) It establishes sodium and potassium gradients across the plasma membrane of all cells.  (2) It helps regulate cell volume.

Vesicular transport occurs when either a material is brought into the cell (endocytosis) or a substance leaves the cell (exocytosis).

Endocytosis is the formation of a vesicle by pinocytosis (cell drinking) or phagocytosis (cell eating).

Such a vesicle will either be degraded by a lysosome or pass across the cell and released.

Exocytosis provides the mechanism for secreting large polar molecules such as hormones and enzymes.

## Membrane Potential (Text page 54)

Membrane potential refers to a separation of opposite charges across the plasma membrane (page 54).

The unequal distribution of a few key ions between the ECF and ICF and their selective movement through the plasma membrane are responsible for the electrical properties of the membrane.

Membrane potential refers to the separation of charges across the membrane.

Because separated charges have the "potential" to do work, a separation  of charges across the membrane is referred to as a membrane potential.

The membrane potential is relatively low therefore the unit used is the millivolt.

Membrane potential is primarily due to differences in distribution and membrane permeability of sodium, potassium and large intracellular anions (page 56).

Cations, sodium and potassium, and large anions are primarily responsible for the slight excess of positive charges outside and the slight excess of negative charges inside the cell.

Sodium is in greater concentration in the extracellular fluid while potassium is in greater concentration in the intracellular fluid.

The large anions are found only inside the cell.

About twenty percent of the membrane potential is directly generated by the sodium potassium pump.

Three sodium ions are transported out for every two potassium ions pumped in.

Most of the pump's role is indirect through its critical contribution to maintaining the concentration gradients directly responsible for the ion movements that generate most of the potential.

Potassium ions and large anions are the pre-dominant ICF  ions.

Sodium and chloride ions are the pre-dominant ECF ions.

Both sodium and potassium respond to the electrical gradient to enter the cell.

Sodium also follows its concentration gradient to enter the cell.

Both ions leak across the membrane.

The sodium-potassium through its direct and indirect mechanisms pump sodium out and potassium in.

The resting membrane potential is -70mV.

The negative sign indicates the inside is negative with respect to the outside of the cell.

The plasma membrane and its various  mechanisms for transporting substance plays a role in homeostasis.

Again the basic cell function is used as a specialization in body systems.

## Nerve and muscle are excitable tissues (page 59).

All cells of the body process a membrane potential related to the distribution and permeability of sodium ions, potassium ions and large intracellular anions.

Nerve and muscle are excitable tissues, capable of producing electrical signals when stimulated.

These fluctuations in membrane potential produce two types of signals: (1) graded potentials (short distance signals) and (2) action potentials (long distance signals).

## Graded potentials die out over short distances (page 60).

Graded potentials are local changes in membrane potential that occur in varying grades or degrees of magnitude; that is the stronger the triggering event, the larger the graded potential.

The triggering event might be (1) a stimulus, (2) an interaction of a chemical messenger, or (3) a spontaneous change of potential caused by imbalances in the leak-pump cycle.

When a graded potential occurs in excitable tissue a different potential exists in this area than in the remainder of the membrane which is still at resting potential.

Because opposite charges attract each other current flows from this active area to the neighboring inactive areas.

Any reduction in resting potential will produce a current flow.

Since these areas on excitable tissues are not insulated the current flow leaks into the ECF.

The graded potential dies out after a short distance.

Graded potentials are critically important to the body as (1) postsynaptic potentials, (2) receptor potentials, (3) end-plate potentials, and (4) pacemaker potentials.

## Action potentials are brief reversals of membrane potential brought about by rapid changes in membrane permeability (page 60).

The rapid reversals of membrane potential in excitable tissues are known as action potentials.

This entire rapid change in potential from resting potential to threshold potential to depolarization and back is called the action potential (also referred to as a spike and firing).

During an action potential, marked changes in membrane permeability of sodium and potassium take place permitting rapid fluxes of these ions.

There are three kinds of channels: (1) Voltage-gated channels, (2) Chemical messenger-gated channels, and (3) mechanically-gated channels.

The rising phase of the action potential (depolarization) is due to a sodium ion influx induced by an explosive increase in sodium ion permeability at threshold.

The falling phase (repolarization) is brought about by a potassium ion efflux caused by a marked increase in potassium ion permeability occurring simultaneously with an inactivation of sodium channels at the peak of the action potential.

## The sodium - potassium pump gradually restores the concentration gradients disrupted by action potentials (page 63).

After the action potential the ion distribution has been altered slightly.

Sodium entered the cell during the rising phase.
Potassium exited the cell during the falling phase.
The sodium - potassium pump restores the normal distribution of these ions.

## Once initiated, action potentials are propagated throughout an excitable cell (page 63).

The nerve cell or neuron has three basic parts: (1) the cell body, (2) the dendrites and (3) the axon.
The nucleus and organelles are in the cell body.
The dendrites serve as antennae to receive signals from other neurons.
The dendrites carry signals toward the cell body.
The axon carries the signal away from the cell body.
Side branches of the axon are called collaterals.
The site of initiation of action potentials is the axon hillock.
The axon hillock is the first part of the axon and the adjacent part of the cell body.
The signal carried by an axon ends at the axon terminal where it stimulates the release of chemical messengers, which excite other cells.
Signal conduction is by local current flow or salutatory conduction.
Once an action potential is initiated in one part of a nerve cell membrane, a self-perpetuating cycle is set up so that the action potential is propagated throughout the rest of the axon automatically in an undiminished fashion.

## Myelination increases the speed of conduction of action potentials (page 66).

The velocity with which an action potential travels down the axon depends on whether or not the fiber is myelinated.
Myelin is composed primarily of lipids.
Water - soluble ions cannot permeate this thick lipid barrier.
Myelin - forming cells wrap around the axon in jelly-roll fashion.
At regular intervals the axon is bare (without myelin) and exposed to the ECF.
These bare areas are known as nodes of Ranvier.
The distance between nodes (about 1mm) is short enough that local current from an active node can reach an adjacent node.
This is saltatory conduction.
Myelinated fibers conduct impulses about fifty times faster than nonmyelinated fibers of comparable diameter.
Multiple sclerosis is a pathophysiological condition in which demyelination of nerve fibers occurs in various locations of the nervous system.
Loss of myelin slows transmission of impulses in the affected neurons.
Scarring associated with myelin damage can injure the underlying axon.

## The refractory period assures unidirectional propagation of the action potential and limits the frequency of action potentials (page 66).

The refractory period has two components: (1) absolute refractory period and (2) relative refractory period.
The absolute refractory period is the time when a recently activated patch of membrane is competely incapable of initiating another action potential no matter how strongly it is stimulated.

The relative refractory period is the time during which a second action potential can be produced only by a stimulus considerably stronger than is usually necessary.
The refractory period ensures the unidirectional propagation of the action potential down the axon away from the initial site of activation.

### Action potentials occur in all-or-none fashion (page 69).

An excitable membrane either responds to a stimulus with a maximal action potential that spreads nondecrementally throughout the membrane, or it does not respond with an action potential at all.
This is known as the all-or-none law.
Weak and strong stimuli evoke the same action potential but stronger stimuli initiate more action potentials.

## Synapses and Neuronal Integration (Text page 69)

### A neurotransmitter carries the signal across a synapse (page 69).

The junction between two neurons is known as the synapse.
The axon terminal of the presynaptic neuron is referred to as a synaptic knob.
The synaptic knob contains synaptic vesicles which store a neurotransmitter.
The space between the presynaptic and postsynaptic membrane is called the synaptic cleft.
The portion of the postsynaptic membrane immediately underlying the synaptic knob is referred to as the subsynaptic membrane.
The action potential in the presynaptic neuron induces the release of the neurotransmitter by opening calcium ion channels in the synaptic knob.
The released neurotransmitter binds to specific sites on the subsynaptic membrane.
This binding triggers the opening of specific ion channels thereby altering the permeability of the postsynaptic neuron.
Synapses operate in one direction only.

### Some synapses excite the postsynaptic neuron whereas others inhibit it (page 71).

At an excitatory synapse the permeability of the sodium and potassium ions is increased by the opening of the respective channels in the subsynaptic membrane.
The result is a net movement of positive ions into the postsynaptic neuron or a small depolarization.
The potential across the membrane is now closer to the threshold potential.
This postsynaptic potential change is called an excitatory postsynaptic potential or EPSP.
At an inhibitory synapse the permeability of the potassium or chloride ion is changed.
The resulting ion movements bring about a small hyperpolarization.
The inside of the postsynaptic neuron becomes more negative.
The difference between this new potential and the threshold potential is greater.
This postsynaptic potential change is called an inhibitory postsynaptic potential or IPSP.
Many chemicals are known or suspected to serve as neurotransmitters.
A given synapse is either always excitatory or always inhibitory.

Neurotransmitters are quickly removed from the synaptic cleft to wipe the post-synaptic slate clean (page 72).

> The transmitter may be inactivated by specific enzymes within the subsynaptic membrane or may be actively taken back up into the axon terminal.
> Once the transmitter is within the synaptic knob it can be stored or destroyed.

The grand postsynaptic potential depends on the sum of the activities of all presynaptic inputs (page 72).

> The grand postsynaptic potential is the total potential produced by the presynaptic neurons.
> The summing of several EPSPs occurring very close together in time because of successive firing of a single presynaptic neuron is known as temporal summation.
> A second method to elicit a grand postsynaptic potential is through concurrent activation of several excitatory inputs or spatial summation.
> Similarly, IPSPs can undergo temporal and spatial summation.

Action potentials are initiated at the axon hillock because it has the lowest threshold (page 74).

> When the summation of EPSPs takes place, the lower threshold of the axon hillock is reached first.

The effectiveness of synaptic transmission can be modified by drugs and disease (page 74).

> Synaptic drugs may act to block an undesirable effect or to enhance a desirable effect.
> Possible drug actions include (1) altering the synthesis, axonal transport, storage, or release of a neurotransmitter; (2) modifying neurotransmitter receptor; (3) influencing neurotransmitter reuptake or destruction; or (4) replacing a deficient neurotransmitter with a substitute transmitter

Neurons are linked to each other through convergence and divergence to form vast and complex nerve pathways (page 76).

> Through convergence a single cell is influenced by thousands of other cells.
> Divergence refers to the branching of axon terminals so that a single cell synapses with many other cells.

Synapses are one of several means by which cells communicate with each other (page 76).

> Communication is critical for the survival of the cells that compose the body.
> Gap junctions are the most intimate means of intercellular communication.
> Some cells utilize signaling molecules on the surface membrane to interact with certain other cells.
> The most common means by which cells communicate is through intercellular chemical messengers: paracrines, neurotransmitters, hormones, and neurohormones.
> Paracrines are local messenger whose effect is only on neighboring cells.
> Neurons communicate with the cells they innervate by releasing neurotransmitters.
> Hormones are long-range chemical messengers released by endocrine glands.
> Neurohormones are messengers released into the blood by neurosecretory neurons.

# KEY TERMS

Action potential-p61
Active transport-p52
All-or-None Law-p69
Concentration gradient-p46
Connexons-p45
Depolarization-p61
Desmosome-p44
Diffusion-p46
Electrical gradient-p48
Excitatory synapse-p71
Facilitated diffusion-p52
Fick's Law of diffusion-p47
Gap junction-p45
Graded potential-p60
Grand postsynaptic potential-p72

Hyperpolarization-p61
Impermeable-p46
Inhibitory synapse-p71
Membrane potential-p54
Membrane-bound enzymes-p40
Nodes of Ranvier-p66
Osmosis-p48
Osmotic pressure-p49
Permeable-p46
Polarization-p61
Postsynaptic neuron-p70
Presynaptic neuron-p70
Receptor sites-p40
Refractory sites-p40
Repolarization-p61

Resting membrane potential-p59
Saltatory conduction-p66
Selectively permeable-p46
Sodium-potassium pump-p53
Spatial summation-p73
State of equilibrium-p47
Steady state-p47
Subsynaptic membrane-p70
Synapse-p69
Synaptic cleft-p70
Synaptic knob-p70
Synaptic vesicles-p70
Temporal summation-p73
Threshold potential-p61
Tight junction-p45

# REVIEW EXERCISES

**True - False** (Answers on page A-5)

_____ 1. Cholesterol is found within the lipid bilayer.

_____ 2. Pinocytosis is the process whereby the white blood cell eats bacteria.

_____ 3. When equilibrium is achieved and no net diffusion is taking place, there is no movement of molecules.

_____ 4. A potential of +30mV is larger than a potential of -70mV.

_____ 5. At resting membrane potential, passive and active forces exactly balance each other so there is no net movement of ions across the membrane.

_____ 6. The survival of every cell depends on the maintenance of intracellular contents unique for that cell type despite the remarkably different composition of the extracellular fluid surrounding it.

_____ 7. All plasma membranes consist mostly of lipids and proteins plus small amounts of carbohydrates.

_____ 8. The small amount of membrane carbohydrate is located only on the inner surface.

_____ 9. The lipid bilayer forms the basic structure of the membrane.

_____ 10. It is **not** possible for a given channel to be open or closed to its specific ion as a result of changes in channel shape in response to a controlling mechanism.

_____ 11. Release of chemicals from nerve cells in response to a nerve impulse is an example of second messenger mechanism.

_____ 12. Different types of cells have different proteins available for phosphorylation and modification by protein kinase.

_____ 13. Interwoven within extracellular matrix gel, there are three types of protein fibers; collagen, elastin, and fibroblasts.

_____ 14. Elastin is a rubberlike protein fiber most abundant in tissues that must be capable of stretching.

_____ 15. The extracellular matrix is secreted by local cells, most commonly by fibronectin.

_____ 16. Sheets of epithelial tissue are joined by gap junctions.

_____ 17. Connexons are formed by the joining of proteins that extend outward from each of the adjacent plasma membranes.

_____ 18. Two properties of particles influence whether they can permeate the plasma membrane without any assistance: The relative solubility of the particle in lipid and the polarity of the particle.

_____ 19. If ten molecules move from A to B while two molecules simultaneously move from B to A, the net diffusion is twelve molecules moving from A to B.

_____ 20. The greater the difference in concentration, the slower the rate of net diffusion.

_____ 21. The more permeable the membrane is to a substance, the more rapidly the substance can diffuse down its concentration gradient.

_____ 22. The larger the surface area available, the greater the rate of diffusion.

_____ 23. The greater the distance, the greater the rate of diffusion.

_____ 24  A solution with a high solute concentration exerts greater osmotic pressure than does a solution with a lower solute concentration.

_____ 25. Cells utilize two different mechanisms to accomplish these selective-transport processes: Carrier-mediated transport and vesicular transport.

_____ 26. If fluid is internalized by endocytosis, the process is termed pinocytosis.

_____ 27. The separation of like charges across the plasma membrane is referred to as a membrane potential.

_____ 28. All living cells have a membrane potential characterized by a slight excess of positive charges outside and a correspondingly slight excess of negative charges on the inside.

_____ 29. Synapses permit two-way transmission of signals between two neurons.

_____ 30. A given synapse may produce EPSPs at one time and IPSPs at another time.

_____ 31. The greater the magnitude of a triggering event such as a stimulus, the smaller the graded potential.

_____ 32. Hyperpolarization means that the membrane returns to resting potential after having been depolarized.

_____ 33. On the device used for recording rapid changes in potential, a decrease in potential is represented as an upward deflection whereas an increase in potential is represented as a downward deflection.

_____ 34. The threshold potential is reached typically between -55mV and -75mV.

_____ 35. The terms action potential, spike, and firing all refer to the same phenomenon of rapid potential reversal.

_____ 36. Membrane channels are formed by proteins that span the thickness of the membrane.

_____ 37. Only about 1 out of 100,000 potassium ions present in the cell leaves during an action potential.

_____ 38. Once an action potential is initiated in one part of a nerve cell membrane, a self-perpetuating cycle is initiated so that the action potential is propagated throughout the rest of the fiber automatically.

_____ 39. Myelinated fibers conduct impulses about 80 times faster than nonmyelinated fibers of comparable size.

_____ 40. As a general rule, the most urgent types of information are transmitted via unmyelinated fibers, whereas the nervous pathways carrying less urgent information are myelinated.

_____ 41. The refractory period ensures the unidirectional propagation of the action potential down the axon from the initial site of activation.

_____ 42. The highest threshold is present at the axon hillock because this region has an abundance of voltage-gated sodium channels.

_____ 43. Parkinson's disease is attributable to an excess of dopamine in a particular region of the brain involved in controlling complex movements.

_____ 44. There are an estimated 100 billion neurons in the brain alone.

**Fill in the Blank** (Answers on page A-6)

45. When _____ occurs, the body's exocrine glands secrete an abnormally thick, sticky mucus.

46. _____ transfer the signal to an intracellular chemical messenger which in turns triggers a preprogrammed series of biochemical events within the cell.

47. _____ brings about the desired intracellular response by opening or closing specific channels in the membrane to regulate the movement of particular ions into or out of the cell.

48. _____ refers to the transfer of a phosphate group from ATP to the protein at the expense of degrading ATP to ADP.

49. The _____ serves as the biological "glue."

50. Once arranged, cells are held together by three different means. They are (1) _____ _____, (2) _____ and (3) _____.

51. List the three major types of protein fibers. (1) _____, (2) _____ and (3) _____.

52. The extracellular matrix is secreted by local cells, most commonly by _____ present in the matrix.

53. Some cells are directly linked together by one of three types of specialized cell junctions, which are (1) _____, (2) _____, and (3) _____.

54. Sheets of epithelial tissue are joined by _____ .
55. _____ forms cablelike fibers or sheets that provide tensile strength.
56. Cells have a _____, which refers to a slight excess of _____ (+/-) charges lined up along the inside of the membrane and separated from a slight excess of positive charges on the outside.
57. _____ have a polar head containing a negatively charged phosphate group and two nonpolar fatty acid tails.
58. A polar end that interacts with water molecules is referred to as _____ and a nonpolar end that will not mix with water is called _____ .
59. _____ contributes to the fluidity as well as the stability of the membrane.
60. Some proteins serves as _____ that transfer substances unable to cross the membrane on their own. Many of the proteins on the outer surface serve as _____ that "recognize" and bind with specific molecules in the environment of the cell.
61. _____ is a rubberlike protein fiber most abundant in tissues that must be capable of easily stretching and then recoiling after the stretching force is removed.
62. _____ are communicating junctions consisting of _____, small connecting tunnels that permit movement of charge-carrying ions between two adjacent cells.
63. _____ promotes cell adhesion and holds cells in position.
64. Gap junctions are found in _____ muscle and _____ muscle.
65. _____ refers to the difference between two opposing movements.
66. When movement of molecules from A to B are exactly matched by movement of molecules from B to A this situation is known as a _____ or _____.
67. If a substance is unable to pass, the membrane is _____ to it.
68. All _____ proteins span the thickness of the plasma membrane and are able to undergo reversible changes in shape so that specific binding sites can alternately be exposed at either side of the membrane.
69. There is a limit to the amount of a substance that can be transported across the membrane via a carrier in a given time. This limit is referred to as _____.
70. With _____, energy is directly required to move a substance uphill.
71. In _____, the plasma membrane surrounds the substance to be ingested, then fuses over the surface, pinching off a membrane-bound vesicle that encloses the engulfed material within the cell.
72. If large multimolecular particles are engulfed, such as bacteria or cellular debris, the process is called _____ .
73. _____ refers to a separation of charges across the membrane or to a difference in the relative number of cations and anions in the ICF and ECF.
74. If a relative difference in charge exists between two adjacent areas, the positive charged ions, called _____, tend to move toward the more _____

charged area, whereas the negatively charged ions, called _____, tend to move toward the more _____ charged area.

75. A difference in charge between two adjacent areas thus produces an _____ _____ that passively induces the movement of ions.

76. The simultaneous existence of an electrical gradient and concentration (chemical) gradient for a particular ion is referred to as an _____.

77. The term _____ refers to the density of the solute (dissolved substance) in a given volume of water.

78. The net diffusion of water is known as _____.

79. The magnitude of opposing pressure necessary to completely stop osmosis is equal to the _____ or hydrostatic pressure.

80. The constant membrane potential that exists when an excitable tissue cell is not displaying rapid changes in potential is referred to as the _____

_____.

81. _____ carry signals toward the cell body. While _____ or _____ _____ carry action potentials away from the cell body.

82. The intervening unmyelinated regions are known as _____ .

83. _____ means the membrane has potential; there is a separation of opposite charges.

84. A faster method of propagation, _____, takes place in myelinated fibers.

85. The space between the presynaptic and postsynaptic neurons is called the _____.

86. The synaptic delay is usually about _____.

87. There are three kinds of channels (1) _____, (2) _____ _____ and (3) _____.

88. The first portion of the axon plus the region of the cell body from which the axon leaves is known as the _____.

89. Name three neurotransmitters that start with the letter G. (1) _____ (2) _____, and (3) _____.

90. Often an action potential is referred to as a _____.

91. The nucleus and organelles are housed in the _____, from which numerous extensions known as _____ typically project like antennae to increase the surface area available for receiving signals from other nerve cells.

92. Myelin is composed primarily of _____.

93. An excitable membrane either responds to a stimulus with a maximal action potential that spreads nondecrementally throughout the membrane, or it does not respond with an action potential at all. This is called the _____.

94. The portion of the postsynaptic membrane immediately underlying the synaptic knob is referred to as the _____.

95. _____ refers to the branching of axon terminals so that a single cell synapses with many other cells.

**Matching:** (Answers on page A-7)

| | | |
|---|---|---|
| _____ | 96. sodium ions | a. diffusion through lipid bilayer |
| _____ | 97. water | b. endocytosis |
| _____ | 98. carbon dioxide | c. exocytosis |
| _____ | 99. calcium ions | d. through protein channel |
| _____ | 100. potassium ions | e. osmosis |
| _____ | 101. oxygen | |
| _____ | 102. bacteria | |
| _____ | 103. chloride ions | |
| _____ | 104. hormones | |
| _____ | 105. fatty acids | |

**Multiple Choice** (Answers on page A-7)

_____ 106. Which of the following acts as "spot rivets" to anchor cells together?
  a. tight junctions
  b. gap junctions
  c. desmosomes
  d. ribosomes
  e. archeosomes

_____ 107. Which type of movement is the net diffusion of water down its own concentration gradient?
  a. carrier-mediated
  b. osmosis
  c. facilitated diffusion
  d. secondary active transport
  e. exocytosis

_____ 108. Which of the following types of cell junctions involve connexons?
  a. tight junctions
  b. gap junctions
  c. desmosomes
  d. all the above
  e. none of the above

_____ 109. Cholesterol contributes to the fluidity and stability of the membrane. Between which molecules is cholesterol found?
  a. membrane proteins
  b. membrane carbohydrates
  c. cell adhesions molecules
  d. phospholipids
  e. G proteins

_____ 110. Phagocytosis is an example of which of the following?
     a. endocytosis
     b. mesocytosis
     c. ectocytosis
     d. hydrocytosis
     e. none of the above

_____ 111. Membrane potential is measured in these units.
     a. volts
     b. nanovolts
     c. kilovolts
     d. millivolts
     e. decivolts

_____ 112. The negative sign of the membrane potential applies to which of the following?
     a. a truly negative potential
     b. the polarity of excess charge on the inside of the membrane
     c. the polarity of excess charge on the outside of the membrane
     d. the concentrations of large anions in the ECF.
     e. none of the above

_____ 113. Multiple sclerosis is a pathophysiological condition in which _____.
     a. the release of the neurotransmitter is blocked
     b. the demyelination of nerve fibers occurs
     c. there is a deficiency of dopamine in a particular region of the brain
     d. the release of gamma-aminobutyrie acid is prevented
     e. the viscosity of mucus increases

_____ 114. Which sequence best describes the events in an action potential?
     a. polarization--hyperpolarization--depolarization--repolarization
     b. hyperpolarization--repolarization--depolarization--polarization
     c. polarization--hyperpolarization--repolarization--depolarization
     d. polarization--depolarization--hyperpolarization--repolarization
     e. depolarization--hyperpolarization--polarization--repolarization

_____ 115. Which of the following releases the neurotransmitters?
     a. axon hillock
     b. dendrite
     c. synaptic vesicles
     d. subsynaptic membrane
     e. synaptic cleft

_____ 116. Which part of the neuron has the lowest threshold potential?
       a. cell body
       b. axon hillock
       c. collaterals
       d. dendrites
       e. axon terminal

_____ 117. Which part of the neuron directs the signal toward the cell body?
       a. nodes of Ranvier
       b. axon hillock
       c. collaterals
       d. dendrites
       e. axon

_____ 118. Which best describes a postsynaptic neuron?  The neuron whose _____
       a. action potentials are propagated toward the synapse.
       b. myelin is formed by oligodendrocytes.
       c. action potentials are propagated away from the synapse.
       d. axons are unmyelinated
       e. dendrites are unmyelinated

_____ 119. Which new drug holds promise in the treatment of Parkinson's disease?
       a. paraquat
       b. meperidine
       c. serotonin
       d. neurotensin
       e. selegiline

**Points to Ponder**
1. After a long hard day you soak in the tub for several minutes.  It feels good  to just relax in the water.  As you dry your body you notice that your fingers and toes are very wrinkled.  What has happened?
2. As you watch people sitting in the smoking section of a restaurant you notice that shortly after one patron lights up a cigarette others soon follow.  What is happening?
3. How would you explain that many different molecules totally man-made, such as some drugs, are transported across membranes?  Even carcinogenic molecules both large and small enter the cell.  How is this possible?
4. Which ion (sodium or potassium) do you think is more important in neuronal physiology?
5. If yelling and grunting enhance an athlete's performance, what do you think happens when the game official tells them to stop yelling and grunting?

**Clinical Perspectives**
1. Why are people with multiple sclerosis likely to lose their ability to control skeletal muscles?
2. Suppose that your doctor is performing neurological tests on your arm using tiny needle electrodes. In doing so he places an electrode adjacent to an axon half-way between the cell body and the axon terminal. When the electrical stimulus is applied which way does the signal travel in the axon?

**Experiment of the Day**
#1        Materials:  2 Glasses
                        1 Face cloth
                        Water
                        Cooking oil
Place one glass half filled with water next to the other one-fourth filled with cooking oil. Carefully insert one end of the dry face cloth into the water glass and the other end into the glass with oil. Be sure both ends are in the respective fluids. Mark the fluid levels. Explain what happens over the next several days.

#2        Demonstrate the following using a set of dominos.
            a.  collateral axon
            b.  convergence
            c.  divergence
            d.  threshold

# CENTRAL NERVOUS SYSTEM  4

## CHAPTER OVERVIEW

To maintain homeostasis, the body, with its many cells, tissues, organs, and systems must be completely aware of any and all changes in the internal and external environment.

The nervous system provides this information using affectors and afferent neurons. The central nervous system, utilizing various components of the brain, processes this information and initiates a response.

Rapid responses are performed by the nervous system through efferent neurons and effectors. The fastest response is the reflex. The endocrine system is used when the response requires duration.

To perform these essential functions the brain is divided, and subdivided, and these divisions interconnected and integrated in such a manner that all parts of the body are controlled.

The brain controls the responses through coordination of the various body systems. In this perspective the nervous system controls homeostasis to preserve the cells of the body.

## CHAPTER OUTLINE

### Introduction (Text Page 83)

The brain is modified in response to environmental influences (page 83).

Many of the basic life-supporting neuronal patterns, such as those controlling respiration and circulation, are similar in all individuals.

Some differences in the nervous systems between individuals are genetically endowed.

Other differences, however, are due to environmental encounters and experiences. Depending on external stimuli and the extent these pathways are used, some are retained, firmly established, and even enhanced, whereas others are eliminated.

## The nervous and endocrine systems have different regulatory responsibilities but share much in common (page 83).

The nervous system coordinates the rapid activities of the body, such as muscle movements.

The endocrine system primarily controls metabolic and other activities that require duration rather than speed, such as maintenance of blood glucose levels.

Both systems ultimately alter their target cells by releasing chemical messengers which interact with specific plasma-membrane proteins of the target cells.

## The nervous system is organized into the central nervous system and the peripheral nervous system (page 83).

The nervous system is organized into the central nervous system (CNS) consisting of the brain and spinal cord, and the peripheral nervous (PNS), consisting of nerve fibers that carry information between the CNS and other parts of the body.

The afferent division of the PNS carries information to the CNS.

Instructions from the CNS are transmitted via the efferent division to effector organs (muscles or glands).

The efferent division is divided into the somatic nervous system which consists of motor neurons that supply the skeletal muscles, and the autonomic nervous system fibers which innervate smooth muscle, cardiac muscle, and glands.

The latter system is subdivided into the sympathetic nervous system and the para-sympathetic nervous system, both of which innervate most of the organs supplied by the autonomic nervous system.

## There are three classes of neurons (page 84).

Three classes of neurons make up the nervous system: afferent neurons, efferent neurons and interneurons.

At its peripheral ending, an afferent neuron has a sensory receptor that generates action potentials in response to a particular type of stimulus.

The afferent neuron cell body is located adjacent to the spinal cord.

A long peripheral axon extends from the receptor to the cell body, and a short central axon passes from the cell body into the spinal cord.

The cell bodies of efferent neurons originate in the CNS.

Efferent axons leave the CNS to course their way to the effector organs they innervate.

Interneurons lie entirely within the CNS.

They serve two main roles: (1) Interneurons lie between the afferent and efferent neurons and are important in the integration of peripheral responses to peripheral information. (2) Interconnections between interneurons themselves are responsible for abstract phenomena associated with the "mind".

## Protection and Nourishment of the Brain (Text page 85)

Neuroglia physically support the interneurons and help sustain them metabolically (page 85).

> About 90% of the cells within the CNS are not neurons but glial cells or neuroglia.
> Glial cells do not initiate or conduct nerve impulses.
> The glial cells serve as connective tissue of the CNS and as such help support the neurons both physically and metabolically.
> The four major types of glial cells are astrocytes, oligodendrocytes, ependymal cells, and microglia.
> Astrocytes provide a number of critical functions: (1) As the main "glue" they hold the neurons together in proper spatial relationships. (2) Astrocytes serve as a scaffold to guide neurons to their proper final destination during fetal brain development. (3) Astrocytes induce the small blood vessels of the brain to undergo the anatomical and functional changes that are responsible for the establishment of the blood brain barrier. (4) They are important in the repair of brain injuries. (5) Astrocytes help support the neurons metabolically. (6) They take up excess potassium ions from the ECF. (7) Finally, astrocytes have receptors for the same neurotransmitters as do neurons.
> Oligodendrocytes form the insulative myelin sheaths around axons in the CNS.
> Ependymal cells line the internal cavities of the CNS.
> The microglia are the phagocytic scavengers of the CNS.
> Unlike neurons, glial cells do not lose the ability to undergo cell division.

The delicate central nervous tissue is well protected (page 88).

> Four major features help protect the CNS: (1) The CNS is enclosed in hard bony structures. (2) Three protective and nourishing membranes, the meninges, lie between the bony covering and the nervous tissue. (3) The brain floats in a special cushioning fluid, the cerebrospinal fluid (CSF). (4) A highly selective blood-brain barrier limits access of blood-borne materials into the brain tissue.

The brain depends on constant delivery of oxygen and glucose by the blood (page 88).

> The brain cannot produce ATP in the absence of oxygen.
> Unlike other tissues, the brain uses only glucose in energy production but does not store any of this nutrient.
> Therefore, the brain is absolutely dependent on a continuous, adequate blood supply of oxygen and glucose.

Brain damage may occur in spite of protective mechanisms (page 89).

> Direct brain damage occurs if the brain is violently shaken or jarred by a forceful impact.
> Further indirect brain damage may occur following a head injury as a consequence of swelling or hemorrhaging within the enclosed confines of the cranium.
> The most common cause of brain damage is not traumatic head injuries, however, but cerebrovascular accidents (strokes).
> When a brain blood vessel ruptures or is blocked by a clot, the brain tissue being supplied by that vessel is deprived of its oxygen and glucose.

Brain damage may also occur as a result of infectious or degenerative neural disorders or brain tumors.

The nature of the ensuing loss of neurological function depends on the area of the brain involved and the extent of permanent damage.

### Headaches are seldom due to brain damage (page 89).

Headaches are the most common form of pain.

Tension, swelling of the mucous membranes, eye disorders, dilation of cerebral blood vessels, increased intracranial pressure, inflammation, and swelling are the most common causes of headaches.

## Cerebral Cortex (Text page 89)

### Newer, more sophisticated regions of the brain are piled on top of older, more primitive regions (page 89).

The brain stem, the oldest and smallest region of the brain, is continuous with the spinal cord.

It controls many of the life-sustaining activities, such as breathing, circulation, and digestion.

Attached at the top rear portion of the brain stem is the cerebellum, which is concerned with maintaining proper position of the body in space and subconscious coordination of motor activity.

On top of the brain stem, tucked within the interior of the cerebrum is the diencephalon.

It houses the hypothalamus, which controls many homeostasis functions, and the thalamus, which performs some primitive sensory processing.

On top of this "cone" of lower brain regions is the cerebrum.

The outer layer of the cerebrum is the cerebral cortex, which caps an inner core that houses the basal nuclei.

The cerebral cortex plays a key role in the most sophisticated neural functions, such as voluntary initiation of movement, final sensory perception, conscious thought, language, personality traits, and other factors we associate with the mind or intellect.

### The cerebral cortex is an outer shell of gray matter covering an inner core of white matter (page 91).

The cerebrum is divided into two halves, the right and left cerebral hemispheres.

They are connected to each other by the corpus callosum.

Each hemisphere is composed of a thin outer shell of gray matter, the cerebral cortex, covering a thick central core of white matter.

Located deep within the white matter is another region of gray matter, the basal nuclei.

Gray matter consists predominately of densely packaged cell bodies and their dendrites as well as glial cells.

Bundles or tracts of myelinated nerve fibers constitute the white matter.

The fiber tracts in the white matter transmit signals from one part of the cerebral cortex to another or between the cortex and other regions of the CNS.

Such communication enables integration between different areas of the cortex and elsewhere.

The four pairs of lobes in the cerebral cortex are specialized for different activities (page 91).

Each half of the cortex is divided into four major lobes: The occipital, temporal, parietal, and frontal lobes.

The occipital lobes, which are located posteriorly are responsible for initially processing visual input.

Sound sensation is initially received by the temporal lobes, located laterally.

The parietal lobes lie to the rear of the central sulcus on each side.

The parietal lobes are primarily responsible for receiving and processing sensory input such as touch, pressure, heat, cold, and pain from the surface of the body.

These sensations are collectively known as somesthetic sensations.

The parietal lobes also perceive awareness of body position, a phenomenon referred to as proprioception.

The somatosensory cortex, the site for initial cortical processing of this somesthetic and proprioceptive input, is located at the front of each parietal lobe immediately behind the central sulcus.

The frontal lobes, lying at the front of the cortex, are responsible for three main functions. (1) Voluntary motor activity. (2) Speaking ability, and (3) Elaboration of thought.

The area at the rear of the frontal lobe immediately in front of the central sulcus and adjacent to the somatosensory cortex is the primary motor cortex.

As in sensory processing, the motor cortex on each side of the brain primarily controls muscles on the opposite side of the body.

Other regions of the nervous system besides the primary motor cortex are important in motor control (page 94).

Lower brain regions and the spinal cord control involuntary skeletal-muscle activity, such as the maintenance of posture.

Although fibers originating from the motor cortex can activate motor neurons to bring about the muscle contraction, the motor cortex itself does not initiate voluntary movement.

The higher motor areas of the brain believed to be involved in the voluntary decision-making period include the supplementary motor area, the premotor cortex and the posterior parietal cortex.

These higher areas all command the primary motor cortex.

Furthermore, the cerebellum appears to play an important role in the planning, initiation, and timing of certain kinds of movement by sending input to the motor areas of the cortex.

movement, such as the opening or closing of the hand.

The premotor cortex located on the lateral surface of each hemisphere in front of the primary motor cortex, is believed to be important in orienting the body and arms toward a specific target.

The premotor cortex is guided by sensory input processed by the posterior parietal cortex, a region that lies posterior to the primary somatosensory cortex.

Language ability has several discrete components controlled by different regions of the cortex (page 95).

> The areas of the brain responsible for language ability are found in only one hemisphere; the left hemisphere in the vast majority of the population.
>
> The primary areas of cortical specialization for language are Broca's and Wernicke's. Broca's area, which is responsible for speaking ability, is located in the left frontal lobe in close association with the motor areas of the cortex that control the muscles necessary for articulation.
>
> Wernicke's area, located in the left cortex at the juncture of the parietal, temporal, and occipital lobes, is concerned with language comprehension.

The association areas of the cerebral cortex are involved in many higher functions (page 96).

> The prefrontal association cortex is the front portion of the frontal lobe just anterior to the premotor cortex.
>
> The roles attributed to this region are (1) planning for voluntary activity; (2) weighing consequences of future actions and choosing between different options for various social or physical situations; and (3) personality traits.
>
> The parietal-temporal-occipital association cortex is found at the interface of the three lobes for which it is named.
>
> In this location, this cortical area pools and integrates somatic, auditory, and visual sensations projected from these three lobes for complex perceptual processing.
>
> This region is also involved in the language pathway connecting Wernicke's area to the visual and auditory cortices.
>
> The limbic association is located on the bottom and adjoining inner portion of each temporal lobe.
>
> This area is concerned primarily with motivation and emotion and is extensively involved in memory.

The cerebral hemispheres have some degree of specialization (page 96).

> The left cerebral hemisphere excels in performance of logical, analytical, sequential, and verbal tasks, such as math, language forms, and philosophy.
>
> The right cerebral hemisphere excels in nonlanguage skills, especially spatial perception and artistic and musical endeavors.

An electroencephalogram is a record of postsynaptic activity in cortical neurons (page 97).

> The electroencephalogram (EEG) has three major uses: (1) It is often used as a clinical tool in the diagnosis of cerebral dysfunction. (2) The EEG is also used to distinguish various stages of sleep. (3) The EEG finds further use in the legal determination of brain death.

## Subcortical Structures and Their Relationship with the Cortex in Higher Brain Functions (Text page 97)

> The subcortical regions include the basal nuclei located in the cerebrum and the thalamus and hypothalamus located in the diencephalon.

The basal nuclei play an important inhibitory role in motor control (page 97).
> The basal nuclei play a complex role in the control of movement in addition to having nonmotor functions that are less well understood.
> The basal nuclei are important in (1) inhibiting muscle tone throughout the body; (2) helping monitor and coordinate slow sustained contractions, especially those related to posture and support; and (3) selecting and maintaining purposeful motor activity while suppressing useless or unwanted patterns of movement.

The thalamus is a sensory relay station and is important in motor control (page 98).
> The diencephalon consists of two main parts, the thalamus and the hypothalamus.
> The thalamus serves as a "relay station" and synaptic integrating center for preliminary processing of all sensory input on its way to the cortex.
> The thalamus, along with the brain stem and cortical association areas, is important in our ability to direct attention to stimuli of interest.

The hypothalamus regulates many homeostatic functions (page 98).
> The hypothalamus is the area of the brain most notably involved in the direct regulation of the internal environment.
> The hypothalamus (1) controls body temperature; (2) controls thirst and urine output; (3) controls food intake; (4) controls anterior pituitary hormone secretion; (5) produces posterior pituitary hormones; (6) controls uterine contractions and milk ejection; (7) serves as a major autonomic nervous system coordinating center, which in turn affect all smooth muscle, cardiac muscle, and exocrine glands; and (8) plays a role in emotional and behavioral patterns.

The limbic system plays a key role in emotion and behavior (page 99).
> The limbic system includes portions of each of the following:   The lobes of the cerebral cortex, the basal nuclei, the thalamus, and the hypothalamus.
> This complex interacting network is associated with emotions, basic survival, and sociosexual behavioral patterns, motivation, and learning.
> Homeostasis drives represent the subjective urges associated with specific bodily needs that motivate appropriate behavior to satisfy those needs.

Norepinephrine, dopamine, and serotonin serve as neurotransmitters in motivated behavioral and emotional pathways (page 100).
> The  underlying neurophysiological mechanisms responsible for the psychological observations of motivated behavior and emotions largely remain a mystery, although the neurotransmitters norepinephrine, dopamine, and serotonin all have been implicated.
> Norepinephrine and dopamine, both chemically classified as catecholamines, are known to be transmitters in the regions that elicit the highest rates of self-stimulation in animals equipped with do-it-yourself devices.

Learning is the acquisition of knowledge as a result of experiences (page 101).
> Learning is the acquisition of knowledge or skills as a consequence of experience, instruction, or both.
> Learning is a change in behavior that occurs as a result of experiences.

Memory is laid down in stages (page 101).

Memory is the storage of acquired knowledge for later recall.

Learning and memory form the basis by which individuals adapt their behavior to their particular external circumstances.

The neural change responsible for retention or storage of knowledge is known as a memory trace.

Concepts, not verbatim information, are generally stored.

Short-term memory lasts for seconds to hours, whereas long-term memory is retained for days to years.

Memory traces are present in multiple regions of the brain (page 102).

The hippocampus, the elongated, medial portion of the temporal lobe that is part of the limbic system, plays a vital role in short-term memory involving the integration of various related stimuli and is also crucial for consolidation into long-term memory.

The hippocampus and the surrounding regions play an important role in declarative memories.

The cerebellum seems to play an essential role in procedural memories involving motor skills gained through repetitive training.

Short-term and long-term memory involve different molecular mechanisms (page 103).

Evidence suggests that short-term memory involves transient modifications in the function of pre-existing synapses, such as a temporary alteration in the amount of neurotransmitter released in response to stimulation within affected nerve pathways.

Long-term memory is believed to involve relatively permanent functional or structural changes between existing neurons in the brain.

# Cerebellum (Text page 103)

The cerebellum is important in balance as well as in planning and execution of voluntary movement (page 103).

The cerebellum, which is attached to the back of the upper portion of the brain stem, lies underneath the occipital lobe of the cortex.

The cerebellum is concerned primarily with subconscious control of motor activity.

The cerebellum (1) contributes to the maintenance of balance, (2) enhances muscle tone, and (3) coordinates skilled, voluntary movements.

The cerebellum and basal nuclei both monitor and adjust motor activity commanded from the motor cortex.

The cerebellum does not have any direct influence on the efferent motor neurons.

The cerebellum helps maintain balance; smoothes out fast, phasic motor activity; and enhances muscle tone.

# Brain Stem (Text page 104)

The brain stem is a vital link between the spinal cord and higher brain regions (page 104).

The brain stem consists of the medulla, pons, and midbrain.

All incoming and outgoing fibers traveling between the periphery and higher brain centers must pass through the brain stem with incoming fibers relaying sensory information to the brain and outgoing fibers carrying command signals from the brain. The functions of the brain stem include the following:

1.  Except for the vagus, the cranial nerves arise from the brain stem and supply structures in the head and neck with both sensory and motor fibers.  Most of the branches of the vagus supply organs  in the thoracic and abdominal cavities.

2.  Collected within the brain stem are neuronal clusters that control heart and blood vessel functions, respiration, and many digestive activities.

3.  The brain stem plays a role in the regulation of muscle reflexes involved with equilibrium and posture.

4.  The reticular formation, found in the brain stem, receives and integrates all synaptic input.  The reticular  activating system (RAS) controls the overall degree of cortical alertness and is important in the ability to direct attention toward specific events.

5. The centers responsible for sleep are also housed within the brain stem.

## Sleep is an active process consisting of alternating periods of slow-wave and para-doxical sleep (page 106).

Consciousness refers to subjective  awareness of the external world and self.

Even though the final level of awareness resides in the cerebral cortex and a crude sense of awareness is detected by the thalamus, conscious experience depends upon the integrated functioning of many parts of the nervous system.

The following states of consciousness are listed in decreasing order of level of arousal; maximum alertness, wakefulness, sleep, and coma.

Sleep is an active process.

There are two types of sleep, characterized by different EEG patterns and different behavior; slow-wave sleep and paradoxical or REM sleep.

Slow-wave sleep occurs in four stages, each displaying progressively slower EEG waves of higher amplitude.

In slow-wave sleep, the person still has considerable muscle tone and frequently shifts body position.

The muscles are completely relaxed in paradoxical sleep with no movement taking place.

Paradoxical sleep is further characterized by rapid eye movement (REM).

The EEG pattern during paradoxical sleep is similar to that of a wide-awake, alert individual.

A person cyclically alternates between the two types of sleep throughout the night.

## The sleep-wake cycle is probably controlled by interactions among three brain stem regions (page 109).

The sleep-wake cycle as well as the various stages of sleep are believed to be due to the cyclical interplay of three different neural systems in the brain stem:  (1) An arousal system, (2) a slow-wave sleep center, and (3) a paradoxical sleep center.

# Spinal Cord (Text page 109)

## The spinal cord extends through the vertebral canal and is connected to the spinal nerves (page 109).

Exiting through the large hole in the base of the skull, the spinal cord is enclosed by the protective vertebral column as it descends through the vertebral canal.

Paired spinal nerves emerge from the spinal cord through spaces formed between the bony, winglike arches of adjacent vertebrae.

There are eight pairs of cervical nerves, twelve thoracic nerves, five lumbar nerves, five sacral nerves, and one coccygeal nerve.

The spinal cord extends only to the level of the first or second lumbar vertebra.

The remaining nerves are greatly elongated in order to exit the vertebral column at their appropriate space.

The gray matter in the spinal cord forms a butterfly-shaped region on the inside and is surrounded by the outer white matter.

The gray matter of the cord consists primarily of neuronal cell bodies and their dendrites, short interneurons, and glial cells.

The white matter is organized into tracts.

Each of these tracts begins or ends within a particular area of the brain.

Some are ascending tracts that transmit to the brain signals derived from afferent input.

Others are descending tracts that relay messages from the brain to the efferent neurons.

Afferent fibers carrying incoming signals enter the spinal cord through the dorsal root.

Efferent fibers carrying outgoing signals leave the spinal cord through the ventral root.

The cell bodies for the afferent neurons at each level are clustered together in a dorsal root ganglion.

The cell bodies of the efferent neurons are in the gray matter of the cord.

The dorsal and ventral roots at each level join to form a spinal nerve.

The spinal nerves and the cranial nerves constitute the peripheral nervous system.

## The spinal cord is responsible for the integration of many basic reflexes (page 111).

The neural pathway involved in accomplishing reflex activity is known as a reflex arc, which typically includes five basic components: A receptor, an afferent pathway, an integrating center, an efferent pathway, and an effector.

There are several types of reflexes involving the reflex arc.

Pathways for unconscious responsiveness digress from the typical reflex arc in two general ways: Responses mediated at least in part by hormones and local responses that do not involve either nerves or hormones.

## KEY TERMS

Afferent neuron-p84
Association areas-p96
Basal nuclei-p97
Dorsal root-p110
Dorsal root ganglion-p110
Efferent neuron-p85
Electroencephalogram-p97
Frontal lobes-p93
Hypothalamus-p98
Interneuron-p85

Learning-p101
Limbic system-p99
Long-term memory-p101
Occipital lobes-p91
Parasympathetic nervous
    system-p84
Parietal lobes-p92
Peripheral nervous
    system-p84

Reflex arc-p111
Short-term memory-p101
Somatic nervous system-p84
Sympathetic nervous
    system-p84
Temporal lobes-p91
Thalamus-p98
Ventral root-p110

# REVIEW EXERCISES

**True-False** (Answers on page A-10)

_____ 1. The brain stem, the oldest and largest region of the brain, is continuous with the spinal cord.

_____ 2. Simple awareness of touch, pressure, or temperature is detected by the thalamus.

_____ 3. The motor cortex on each side of the brain primarily controls muscles on the same side of the body.

_____ 4. The cerebellum appears to play an important role in the planning, initiation, and timing of certain kinds of movement by sending input to the motor areas of the cortex.

_____ 5. The areas of the brain responsible for language ability are found in only one hemisphere--the right hemisphere.

_____ 6. Epileptic seizures occur when a large collection of neurons abnormally undergo synchronous action potentials that produce stereotypical, involuntary spasms and alterations in behavior.

_____ 7. The central nervous system, (CNS), consist of the brain and the spinal cord.

_____ 8. The efferent division carries information to the CNS apprising it of the external environment and providing status reports on internal activities being regulated by the nervous system.

_____ 9. Afferent neurons are shaped differently than efferent neurons and interneurons.

_____ 10. The efferent neuron cell body, is located adjacent to the spinal cord.

_____ 11. The cell bodies of afferent neurons originate in the CNS.

_____ 12. Interneurons lie between the afferent and efferent neurons.

_____ 13. Interconnections between interneurons are responsible for the abstract phenomena associated with the mind.

_____ 14. The cerebellum essentially acts as "middle management," comparing the "intentions" or "orders" of the higher centers with the "performance" of the muscles and then correcting any "errors" or deviations from the intended movement.

_____ 15. The cerebellum does not have any direct influence on the efferent motor neurons.

_____ 16. The cerebellum plays a role in the planning and initiation of voluntary activity by providing input to the cortical motor areas.

_____ 17. The cerebellum helps maintain balance; smoothes out fast, phasic motor activity; and enhances muscle tone.

_____ 18. About ninety percent of the cells within the CNS are neurons.

_____ 19. Glial cells do not initiate or conduct nerve impulses.

_____ 20. Astrocytes form myelin sheaths in CNS.

_____ 21. Ependymal cells line the internal cavities of brain and spinal cord.

_____ 22. The ependymal cell lining of the ventricles contributes to the formation of cerebrospinal fluid.

_____ 23. The blood-brain barrier protects the delicate brain and spinal cord from chemical fluctuations in the blood and minimizes the possibility that potentially harmful blood-borne substances might reach the central neural tissue.

_____ 24. The brain can protect ATP in the absence of oxygen.

_____ 25. Damaged brain cells release excessive amounts of glutamate.

_____ 26. The spinal cord is about 28 inches long.

_____ 27. There are eight pairs of cervical nerves, 12 thoracic nerves, 7 lumbar nerves, 5 sacral nerves, and 1 coccygeal nerve.

_____ 28. Ascending tracts relay messages from the brain to efferent neurons.

_____ 29. Efferent fibers carrying incoming signals enter the spinal cord through the dorsal root.

_____ 30. The cell bodies for the efferent neurons originate in the gray matter and send axons out through the ventral root.

_____ 31. The thirty-one pairs of spinal nerves, along with the twelve pairs of cranial nerves that arise from the brain, constitute the peripheral nervous system.

_____ 32. All incoming and outgoing fibers traversing between the periphery and higher brain centers must pass through the brain stem.

_____ 33. Most of the branches of the vagus nerve supply organs in the thoracic and abdominal cavities.

_____ 34. The following states of consciousness are listed in increasing order of level of arousal:  Coma, sleep, wakefulness, and maximum alertness.

_____ 35. The brains overall level of activity is reduced during sleep.

_____ 36. Paradoxical sleep occupies eighty percent of total sleeping time throughout adolescence and most of adulthood.

_____ 37. Slow-wave sleep is characterized by REM.

_____ 38. The subcortical regions include the basal nuclei located in the cerebrum and the thalamus and hypothalamus located in the diencephalon.

_____ 39. The thalamus exerts an inhibitory effect on motor activity by acting through neurons in the brain.

_____ 40. The hypothalamus is the area of the brain most notably involved in the direct regulation of the internal environment.

_____ 41. Short-term memory is believed to involve relatively permanent functional or structural changes between existing neurons in the brain.

_____ 42. Numerous hormones and neuropeptides are known to affect learning and memory processes.

_____ 43. Both the nervous and the endocrine system ultimately alter their target cells by releasing chemical messengers.

_____ 44. Both the nervous and the endocrine system has an influence on other major control systems.

_____ 45. The nervous system has a very slow speed of response.

_____ 46. Binding of a hormone with target-cell receptors initiates a reaction (or series of reactions) that culminates in the hormone's final effect.

_____ 47. The endocrine system is responsible for coordinating rapid, precise responses.

_____ 48. The nervous system directly or indirectly controls the secretion of many hormones.

_____ 49. In the nervous system, neurotransmitters are released into the synaptic knob.

**Fill in the Blanks** (Answers on page A-11)

50. The _____ regulates muscle tone and coordinates skilled, voluntary movements.

51. The _____ is important for the maintenance of balance and control of eye movement.

52. The _____ plays a role in the planning and initiation of voluntary activity by providing input to the cortical motor areas.

53. There are two types of reflexes: _____ and _____.

54. The spinal nerves are named according to the region of the vertebral column from which they emerge:  There are eight pairs of _____ nerves, twelve _____ nerves, five _____ nerves, five _____ nerves, and one coccygeal nerve.

55. The thick bundle of elongated nerve roots within the lower vertebral canal is known as the _____.

56. _____ fibers carrying incoming signals enter the spinal cord through the dorsal root; _____ fibers carrying outgoing signals leave through the ventral root.

57. A collection of neuronal cell bodies located outside of the CNS is called a _____, whereas a functional collection of cell bodies within the CNS is referred to as a

_____.

58. The dorsal and ventral roots at each level join to form a _____.

59. A _____ is a bundle of peripheral neuronal axons, some afferent and some efferent, which are enclosed by a connective-tissue covering.

60. The diencephalon houses two brain components:  (1) _____ and (2) _____.

61. The _____ is the largest portion of the human brain.

62. Each hemisphere is composed of a thin outer shell of gray matter called _____.

63. _____ matter consists predominantly of densely packaged cell bodies and their dendrites as well as glial cells.  Bundles or tracts of myelinated nerve fibers (axons) constitute the _____ matter.

64. The _____ are primarily responsible for receiving and processing sensory input such as touch, pressure, heat, cold, and pain from the surface of the body.  These sensations are collectively known as _____.

65. _____ area, which is responsible for speaking ability, is located in the left _____. _____ area, located in the left _____ at the juncture of the parietal, temporal, and occipital lobes, is concerned with language comprehension.

66. About ninety percent of the cells within the CNS are _____ or _____.

67. _____ serve as a scaffold to guide neurons to their proper final destination during fetal brain development.

68. _____ form the insulative myelin sheaths around axons in the CNS.

69. _____ phagocytic cells that are the scavengers of the CNS.

70. The brain stem consists of the _____, _____, and _____.

71. Running throughout the entire brain stem and into the thalamus is a widespread network of interconnected neurons called the _____.

72. _____ refers to subjective awareness of the external world and self.

73. _____ refers to the total unresponsiveness of a living person to external stimuli, caused be either by brain stem damage or by widespread depression of the cerebral cortex.

74. _____ sleep can be considered either the deepest sleep or the lightest sleep.

75. In _____ sleep a person still has considerable muscle tone and frequently shifts body position.

76. In _____ sleep, muscles are completely relaxed with no movement taking place.

77. The subcortical regions include the _____ located in the cerebrum and the _____ and _____ located in the diencephalon.

78. The _____ serves as a "relay station" and synaptic integrating center for preliminary processing of all sensory input on its way to the cortex.

79. The _____ controls body temperature, controls food intake and controls thirst.

80. _____ is the ability to direct behavior toward specific goals.

81. _____ is the acquisition of knowledge or skills as a consequence of experience, instruction, or both, and is a change in behavior that occurs as a result of experiences.

82. The process of transferring and fixing short-term memory traces into long-term memory stores is known as _____.

83. The nervous system is organized into the central nervous system, consisting of the _____ and _____, and the _____.

84. The _____ carries information to the CNS.

85. Three classes of neurons make up the nervous system. (1) _____, (2) _____, and (3) _____.

86. Interneurons lie entirely within the _____.

**Matching**: Match function to type of glial cell. (Answers on page A-12)

_____ 87. Contribute to formation of cerebrospinal fluid

_____ 88. Form myelin sheaths in CNS

_____ 89. Induce formation of blood-brain barrier

_____ 90. Serve as scaffold during fetal brain development

_____ 91. Take up excess potassium to help maintain proper brain ECF ion concentration and normal neural excitability

_____ 92. Physically support neurons in proper spatial relationship

_____ 93. Play a role in defense of brain as phagocytic scavengers

_____ 94. Form neural scar tissue

_____ 95. Line internal cavities of brain and spinal cord

_____ 96. Take up and degrade released neurotransmitters into raw materials for synthesis of more neuro-transmitters by neurons

_____ 97. Possess receptors for neurotransmitters, which may be important in a chemical signaling system

a. Astrocytes
b. Ependymal cells
c. Microglia
d. Oligodendrocytes

**Multiple Choice** (Answers on page A-12)

_____ 98. Which of the nerve cells loses its ability to undergo cell division?
   a. astrocytes
   b. ependymal cells
   c. interneurons
   d. oligodendrocytes
   e. microglia

_____ 99. Which of the following is divided into the parasympathetic nervous system?
   a. limbic system
   b. somatic nervous system
   c. autonomic nervous system
   d. afferent nervous system
   e. CNS

_____ 100. Which type of sleep is characterized by having progressively slower EEG waves of higher amplitude?
   a. REM sleep
   b. slow-wave sleep
   c. paradoxical sleep
   d. rhythmic sleep
   e. parenthetical sleep

_____ 101. This disease is characterized early as only short-term memory loss, as the disease progresses, even firmly entrenched long-term memories is impaired.
   a. cerebrovascular accident
   b. dyslexia
   c. aphasias
   d. Alzheimer's
   e. Parkinson's

_____ 102. Which of the following is an inability to store memory in long-term storage?
   a. Alzheimer's disease
   b. anterograde amnesia
   c. aphasia
   d. retrograde amnesia
   e. Parkinson's disease

_____ 103. Which of the following is a collection of neuronal cell bodies located outside the CNS?
   a. center
   b. nucleus
   c. ganglion
   d. tract
   e. root

_____ 104. Which disorder is known as a lazy eye?
   a. amblyopia
   b. dyslexia
   c. aphasia
   d. Parkinson's disease
   e. Schizophrenia

_____ 105. Which of the following is a language disorder caused by damage to specific cortical areas?
   a. amblyopia
   b. hydrocephalus
   c. Alzheimer's disease
   d. Schizophrenia
   e. aphasia

_____ 106. This disorder is diagnosed when a large collection of neurons abnormally undergo synchronous action potentials that produce stereotypical involuntary spasms and alterations in behavior.
   a. Alzheimer's disease
   b. Schizophrenia
   c. Parkinson's disease
   d. epilepsy
   e. dyslexia

_____ 107. Which area of the brain produces posterior pituitary hormones?
   a. hippocampus
   b. basal nuclei
   c. hypothalamus
   d. corpus callosum
   e. thalamus

## Points to Ponder

1. Suppose you had a dream last night in which you, David Crockett and Columbus were building a space laboratory on the moon. How did your brain come up with such a story?
2. After playing tennis yesterday you stopped at a fast-food store for something to drink and then decided to have a snack. What were the cues your brain received and how were they processed?
3. It has been demonstrated by experimentation that learning can occur by rewarding the animal if it responds in the desirable manner. If the desirable response is for the rat to run through a maze, and we reward the animal with food upon completion of the task, would it be correct to assume that learning occurs only if the rat is stressed? What happens if you feed the rat first?

## Clinical Perspectives

1. What would you suggest to a person complaining of frequent severe headaches of long duration?
2. What loss in body function would you expect in a car-crash victim with extensive injury to the cerebellum, the left temporal lobe, and the left occipital lobe?

# PERIPHERAL NERVOUS SYSTEM 5

## CHAPTER OVERVIEW

Perception plays a major role in maintaining homeostasis. Any change in the environment must be dealt with by the organism if homeostasis is to be maintained. The necessary internal and external environmental information is picked up at the peripheral endings of afferent neurons. These specialized endings are classified as follows: Photoreceptors, Mechanoreceptors, Thermoreceptors, Osmoreceptors, Chemoreceptors, and Nociceptors. All of these receptors depolarize or hyper-polarize to produce a receptor or generator potential that is a graded potential. Graded potentials initiate action potentials thus the peripheral nervous system delivers the signal to the CNS for interpretation. Each type receptor is part of a designed pathway going to a specific part of the CNS. With the information reaching the CNS, the nervous system performs its all important role of maintaining homeostasis. In this chapter special emphasis is placed on pain (Nociceptors), Vision (Photoreceptor), Hearing and Equilibrium (Mechanoreceptors), and Taste and Smell (Chemoreceptors). As one of the major control systems of the body, the CNS transmits impulses through the efferent portion of the peripheral nervous system to the effector organs.
The transmitted signal is usually the result of a CNS interpretation of conditions that warrant changes in order to maintain homeostasis. Voluntary and involuntary effector organs are controlled using the somatic and autonomic nervous system respectively.
The key to this very precise autonomic control is in the neurotransmitters, acetylcholine and norepinephrine, and the dual innervation by the sympathetic and parasympathetic systems. The somatic nervous system acts through the release of a neurotransmitter at the neuromuscular junction to cause skeletal-muscle contractions.
The efferent division of the peripheral nervous system is the final link in the CNS control of the body.
The actions elicited by the efferent neurons are essential for homeostasis.

## CHAPTER OUTLINE

### Introduction:  Afferent Input (Text page 118)

The afferent division of the peripheral nervous system sends information about the internal and external environment to the CNS.

Afferent input that does not reach the level of conscious awareness includes internal information , such as blood pressure.

Afferent input that does reach the level of conscious awareness is called sensory input.

Perception is our conscious interpretation of the external world as created by the brain from a pattern of nerve impulses delivered to it from sensory receptors.

### Receptor Physiology (Text page 118)

#### Receptors have differential sensitivities to various stimuli (page 118).

At their peripheral endings afferent neurons have receptors that apprise the CNS of detectable changes in both the external world and the internal environment by generating action potentials in response to the stimuli.

Receptors must convert other forms of energy into action potentials.

This energy conversion process is known as transduction.

Photoreceptors are responsive to light.

Mechanoreceptors are sensitive to mechanical energy.

Thermoreceptors are sensitive to heat and cold.

Osmoreceptors detect changes in the concentration of solutes in the body fluids and the resultant changes in osmotic activity.

Chemoreceptors are sensitive to specific chemicals.

Nociceptors or pain receptors are sensitive to tissue damage such as pinching or burning or to distortion of tissue.

Intense stimulation of any receptor is also perceived as painful.

#### Altered membrane permeability of receptors in response to a stimuli produces a graded receptor potential (page 120).

A receptor may be either a specialized ending of the afferent neuron or a separate cell closely associated with the peripheral endings of the neuron.

The receptor potential is a graded potential whose amplitude and duration can vary, depending on the strength and the rate of application or removal of the stimulus.

A larger receptor potential cannot bring about a larger action potential (all-or-none law), but it can induce more rapid firing of action potentials.

Stimulus intensity is distinguished both by the frequency of action potentials generated in the afferent neuron (frequency code) and by the number of receptors activated within the area (population code).

#### Receptors may adapt slowly or rapidly to sustained stimulation (page 120).

There are two types of receptors--tonic receptors and phasic receptors--based on their speed of adaptation.

Tonic receptors do not adapt at all or adapt slowly.

Phasic receptors are rapidly adapting receptors.

Phasic receptors are useful in situations where it is important to signal a change in stimulus intensity rather than to relay status quo information.

Each somatosensory pathway is "labeled" according to modality and location (page 121).

> On reaching the spinal cord, afferent information has two possible destinies: (1) it may become part of a reflex arc, bringing about an appropriate effector response, or (2) may be relayed upward to the brain via ascending pathways for further processing and possible conscious awareness.
>
> A particular sensory input is "projected" to a specific region of the cortex.
>
> Information is kept separated within specific labeled lines between the periphery and the cortex.
>
> In this way, even though all information is propagated to the CNS via the same type of signal, the brain can decode the type and location of the stimulus.

Acuity is influenced by receptive field size (page 121).

> Each sensory neuron responds to stimulus information only within a circumscribed region of the skin surface surrounding it; this region is known as its receptive field.
>
> The smaller the receptive field in a region, the greater its acuity or discriminative ability.
>
> More cortical space is allotted for sensory reception from areas with smaller receptive fields and, accordingly, greater tactile discriminative ability.

# Pain (Text page 122)

Stimulation of nociceptors elicits the perception of pain plus motivational and emotional responses (page 122).

> Pain is primarily a protective mechanism meant to bring to conscious awareness the fact that tissue damage is occurring or is about to occur.
>
> There are three categories of pain receptors: mechanical nociceptors respond to mechanical damage such as cutting, crushing, or pinching; thermal nociceptors respond to temperature extremes; and polymodal nociceptors respond equally to all kinds of damaging stimuli, including irritating chemicals released from injured tissues.
>
> None of the nociceptors have specialized receptor structures.
>
> Nociceptors do not adapt to sustained or repetitive stimulation.
>
> All nociceptors can be sensitized by the presence of prostaglandins.
>
> Signals arising from mechanical and thermal nociceptors are transmitted over large myelinated fibers (the fast pain pathway).
>
> Impulses from polymodal nociceptors are carried by small unmyelinated fibers (slow pain pathway).

The brain has a built-in analgesic system (page 123).

> In addition to the chain of neurons connecting peripheral nociceptors with higher CNS structures for pain perception, the CNS also contains a neuronal system that suppresses pain; a built-in analgesic system.
>
> Electrical stimulation of the periaqueductal gray matter results in profound analgesia, as does stimulation of the reticular formation.
>
> These two regions are thought to be part of a descending analgesic pathway that blocks the release of substance P from afferent pain fiber terminals.
>
> Endogenous opiates serve as analgesic neurotransmitters; they are released from a descending analgesic pathway and bind with opiate receptors on the afferent pre-synaptic terminal.
>
> This binding suppresses the release of substance P.

## Eye:  Vision (Text page 125)

### The eye is a fluid-filled sphere enclosed by three specialized tissue layers (page 125).

Most of the eyeball is covered by a tough outer layer of connective tissue, the sclera.
Anteriorly, the outer layer consists of the transparent cornea through which light rays pass into the interior of the eye.
The middle layer underneath the sclera is the highly pigmented choroid.
The choroid layer becomes specialized anteriorly to form the ciliary body and the iris.
The innermost coat under the choroid is the retina, which consists of an outer pigmented layer and an inner nervous-tissue layer.
The retina contains the rods and cones.
The anterior cavity between the cornea and lens contains a clear watery fluid, the aqueous humor, and the larger posterior cavity between the lens and the retina contains a semifluid, jellylike substance, the vitreous humor.

### The amount of light entering the eye is controlled by the iris (page 127).

The iris is a thin pigmented smooth muscle that forms a visible ringlike structure within the aqueous humor.
The round opening in the center of the iris is the pupil.
The size of the pupil is adjusted by the iris muscles.
Iris muscles are controlled by the autonomic nervous system.

### The eye refracts the entering light to focus the image on the retina (page 127).

The bending of a light ray (refraction) occurs when the ray passes from a medium of one density into a medium of a different density.
A lens with a convex surface converges light rays, bringing them closer together, a requirement for bringing an image to a focal point.
Refractive surfaces of the eye are convex.
The two structures most important in the eye's refractive ability are the cornea and the lens.

### Accommodation increases the strength of the lens for near vision (page 129).

The ability to adjust the strength of the lens so that both near and far sources can be focused on the retina is known as accommodation.
The strength of the lens depends on its shape, which in turn is regulated by the ciliary muscle.
The ciliary body has two major components:  The ciliary muscle and the capillary network that produces the aqueous humor.

### Light must pass through several retinal layers before reaching the photoreceptors (page 130).

The major function of the eye is to focus light rays from the environment on the rods and cones, the photoreceptors cells of the retina.
The photoreceptors then transform the light energy into electrical signals for transmission to the CNS.
The neural portion of the retina consists of three layers:  (1) The outermost layer containing the rods and cones, whose light-sensitive ends face the choroid; (2) a middle

layer of bipolar neurons; and (3) an inner layer of ganglion cells.

Axon of the ganglion cells join together to form the optic nerve.

The point on the retina at which the optic nerve leaves and through which blood  vessels pass is the optic disc.

In the fovea, which is located in the exact center of the retina, the bipolar and ganglion cell layers are pulled aside so that light directly strikes the photoreceptors.

This feature, coupled with the fact that only cones are found here, makes the fovea the point of most distinct vision.

## Phototransduction by retinal cells converts light stimuli into neural signals that are perceived by the visual cortex (page 131).

Photoreceptors consists of three parts:  (1) an outer segment, which lies closest to the eye's exterior, facing the choroid, and detects the light stimulus; (2) an inner segment, which contains the metabolic mechinery; and (3) a synaptic terminal, which lies closest to the eye's interior, facing the bipolar neurons and transmits the signal generated in the photoreceptor upon light stimulation to the next cells in the visual pathway.

The outer segment is rod-shaped in rods and cone-shaped in cones.

The photopigment which is found in the outer segment, consists of an enzymatic protein called opsin combined with retinene, a derivative of vitamin A.

There are four different photopigments, one in rods and one in each of three types of cones.

Rhodopsin, the rod photopigment, cannot discriminate between various wavelengths.

Rods provide vision only in shades of gray by detecting different intensities.

The photopigments in the three types of cones--red, green, and blue cones-- respond selectively to various wavelengths, making color vision possible.

The photoreceptors synapse with bipolar cells.

These cells terminate on the ganglion cells, whose axons form the optic nerve for transmission of signals to the brain.

Bipolar cells display graded potentials similar to the photoreceptors.

Action potentials do not originate until the ganglion cells, the first neurons in the chain that must propagate the visual message to the brain.

## Rods provide indistinct gray vision at night, whereas cones provide sharp color vision during the day (page 135).

Cones provide sharp vision with high resolution for fine detail.

Rods have low acuity but high sensitivity, so they respond to the dim light of night.

There is little convergence of neurons in the retinal pathway for cone output.

In contrast, there is much convergence of neurons in rod pathways.

## The sensitivity of the eyes can vary markedly through dark and light adaptation (page 135).

When you go from bright sunlight into darkened surroundings, you cannot see anything at first, but gradually you begin to distinguish objects as a result of the process of dark adaptation.

When you move from dark to the light, your eyes are very sensitive to the light at first.

The sensitivity of the eyes decreases and normal contrasts can once again be detected, a process known as light adaptation.

Color vision depends on the ratios of stimulation of the three cone types (page 136).

>Our perception of the many colors of the world depends on the three cone types various ratios of stimulation in response to different wave lengths.

Visual information is separated within the visual pathway before it is integrated into a perceptual image of the visual field by the cortex (page 137).

>The field of view that can be seen without moving the head is known as the visual field.
>The overlapping area seen by both eyes at the same time is known as the binocular field of vision, which is important for depth perception.
>Because of the pattern of wiring between the eyes and the visual cortex, the left half of the cortex receives information only from the right half of the visual field, and the right half receives input only from the left half of the visual field of both eyes.

Protective mechanisms help prevent eye injuries (page 137).

>Several mechanisms help protect the eyes from injury; bony sockets, eyelids, tears, and eyelashes.

## Ear:  Hearing and Equilibrium (Text page 138)

Sound waves consist of alternate regions of compression and rarefaction of air molecules (page 140).

>Hearing is the neural perception of sound energy.
>Sound waves are traveling vibrations of air that consist of regions of high pressure caused by compression of air molecules alternating with regions of low pressure caused by rarefaction of the molecules.
>Sound waves can also travel through media other than air, such as water.
>Sound is characterized by its pitch, intensity, and timbre.

The external and middle ear convert airborne sound waves into fluid vibrations in the inner ear (page 141).

>The pinna collects sound waves and channels them down the external ear canal.
>The tympanic membrane, which is stretched across the entrance to the middle ear, vibrates when struck by sound waves.
>The middle ear transfers the vibratory movements of the tympanic membrane to the fluid of the inner ear.
>This transfer is facilitated by a moveable chain of three small bones (the malleus, incus, and stapes) that extend across the middle ear.
>The stapes is attached to the oval window, the entrance into the fluid-filled cochlear.

Hair cells in the organ of Corti transduce fluid movements into neural signals (page 143).

>The snail-shaped cochlear portion of the inner ear is a coiled tubular system lying deep within the temporal bone.
>The cochlea is divided throughout most of its length into three fluid-filled longitudinal compartments.
>The round window seals the middle ear.

The basilar membrane forms the floor of the cochlear duct and bears the organ of Corti, the sense organ for hearing.
The organ of Corti contains hair cells that are the receptors of sound.
Pressure waves through the basilar membrane cause this membrane to vibrate in synchrony with the pressure wave.
Since the organ of Corti rides on the basilar membrane, the hair cells also move up and down as the basilar membrane oscillates.
This back and forth mechanical deformation of the hairs brings about potential changes--the receptor potential--at the same frequency as the original sound stimulus.
The receptor potential is converted into action potentials in the afferent nerve fibers that make up the auditory nerve.

## Pitch discrimination depends on the region of the basilar membrane that vibrates; loudness discrimination depends on the amplitude of the vibration (page 143).

Different regions of the basilar membranes naturally vibrate maximally at different frequencies; that is, each frequency displays peak vibration at a different position along the membrane.
Intensity discrimination depends on the amplitude of the vibration.
A greater tympanic membrane deflection is converted into greater amplitude of basilar membrane movement in the region of peak responsiveness.
The CNS interprets greater basilar membrane oscillation as louder sound.

## Deafness is caused by defects either in conduction or neural processing of sound waves (page 145).

Conduction deafness occurs when sound waves are not adequately conducted through the external and middle portions of the ear to set the fluids in the inner ear in motion.
In sensorineural deafness, the sound waves are transmitted to the inner ear, but they are not translated into nerve signals that are interpreted by the brain as sound sensations.

## The vestibular apparatus detects position and motion of the head and is important for equilibrium and coordination of head, eye, and body movements (page 145).

The vestibular apparatus consists of two sets of structures within the temporal bone near the cochlea; the semicircular canals and the otolith organs.
The semicircular canals detect rotational and angular acceleration or deceleration of the head.
Each ear contains three semicircular canals arranged three-dimensionally in planes that lie at right angles to each other.
The receptive hair cells of each semicircular canal are situated on top of a ridge.
The hairs are embedded in the cupula.
As the head starts to move, the fluid within the semicircular canal does not initially move in the direction of the rotation but lags behind because of its inertia.
The otolith organs provide information about the position of the head relative to gravity and detect changes in rate of linear motion.
The utricle and saccule are saclike structures housed within a bony chamber situated between the semicircular canals and the cochlea.
The hairs of the receptive hair cells in these sense organs also protrude into an overlying gelatinous sheet, whose movement displaces the hairs and results in changes in hair cell potential.

When a person is in an upright position, the hairs within the utricles are oriented vertically and the saccule hairs are lined up horizontally.
Signals are carried to the brain stem and to the cerebellum.
Here the vestibular information is integrated with input from the skin surface, eyes, joints and muscles for: (1) maintaining balance and desired posture; (2) controlling the external eye muscles so that the eyes remain fixed on the same point; and (3) perceiving motion and orientation.

## Chemical Senses:  Taste and Smell (Text page 149)

Taste sensation is coded by patterns of activity in various taste bud receptors (page 150).

The chemoreceptors for taste sensation are packaged in taste buds.
Each taste bud has a small opening, the taste pore.
Taste receptor cells are modified epithelial cells with microvilli.
The plasma membrane of the microvilli contains receptor sites that bind selectively with chemical molecules in the environment.
Binding of a taste-provoking chemical with a receptor cell alters its ionic channels to produces a receptor potential.
The receptor potential initiates an action potential afferent nerve fibers.
Signals in these sensory inputs are conveyed  to the cortical gustatory area.
All tastes are varying combinations of four primary tastes: salty, sour, sweet, and bitter.

Smell is the least understood of the special senses (page 151).

To be smelled, a substance must be (1) sufficiently volatile that some of its molecules can enter the nose and (2) sufficiently water-soluble that it can dissolve in the mucus layer coating the olfactory mucosa.
Binding of an odoriferous molecule to a specialized attachment site on the cilia brings about a receptor potential.
The receptor potential generates the action potential in the afferent fiber.
The afferent fibers pass through the bone and immediately synapse in the olfactory bulb.
Fibers leaving the olfactory bulb travel in two different routes: (1) a subcortical route going to regions of the limbic system, and (2) a thalamic - cortical route.

## Introduction:  Efferent Output (Text page 152)

The efferent division of the peripheral nervous system is the communication link by which the CNS controls the activities of muscles and glands.
Cardiac muscle, smooth muscle, most exocrine glands, and some endocrine glands are innervated by the autonomic nervous system.
Skeletal muscle is innervated by the somatic nervous system.
Two neurotransmitters--acetylcholine and norepinephrineare released from efferent neuronal terminals to elicit essentially all the neurally controlled effector organ responses.

## Autonomic Nervous System (Text page 153)

An autonomic nerve pathway consists of a two-neuron chain, with the terminal neurotransmitter differing between sympathetic and parasympathetic nerves (page 153).

Each autonomic nerve pathway consists of a two-neuron chain.

The cell body of the first neuron is located in the CNS.

Its axon, the preganglionic fiber, synapses with the cell body of the second neuron, which lies within a ganglion outside the CNS.

The axon of the second neuron, the postganglionic fiber, innervates the effector organ.

The autonomic nervous system consists of two divisions--the sympathetic and the parasympathetic nervous system.

Sympathetic nerve fibers originate in the thoracic and lumbar regions of the spinal cord.

Most sympathetic preganglionic fibers are very short, synapsing within a sympathetic ganglion chain located along either side of the spinal cord.

Long postganglionic fibers terminate on the effector organs.

Parasympathetic preganglionic fibers arise from the cranial and sacral areas of the CNS. These fibers are long because they do not end until they reach terminal ganglia that lie in or near the effector organs.

Parasympathetic postganglionic fibers are very short.

Sympathetic and parasympathetic preganglionic fibers release the same neurotransmitter, acetylcholine (ACh).

Parasympathetic postganglionic fibers release acetylcholine and are called cholinergic fibers.

Most sympathetic postganglionic fibers are called adrenergic fibers because they release norepinephrine.

The terminal branches of autonomic fibers contain numerous swellings, or varicosites, that simultaneously release neurotransmitter over a large area of the innervated organ.

This diffuse release means whole organs instead of discrete cells are typically influenced by autonomic activity.

## The autonomic nervous system controls involuntary visceral organ activities (page 153).

The autonomic nervous system regulates visceral activities normally outside the realm of consciousness and voluntary control, such as circulation, digestion, sweating and pupillary size.

## The sympathetic and parasympathetic nervous systems dually innervate most visceral organs (page 155).

Most visceral organs are innervated by both sympathetic and parasympathetic nerve fibers.

Both systems increase the activity of some organs and reduce the activity of others.

Normally some level of action potential activity exists in both, the sympathetic and the parasympathetic fibers supplying a particular organ.

This ongoing activity is called sympathetic or parasympathetic tone or tonic activity.

The sympathetic system promotes responses that prepare the body for strenuous physical activity in the face of emergency or stressful situations.

This response is typically referred to as a fight-or-flight response.

Dual innervation enables precise control over an organs activity.

The adrenal medulla, an endocrine gland, is a modified part of the sympathetic nervous system (page 157).

> The adrenal medulla is considered to be a modified sympathetic ganglion that does not give rise to postganglionic fibers.
> Upon stimulation by the preganglionic fiber the adrenal medulla secretes hormones into the blood.
> The hormones are identical or similar to postganglionic sympathetic neurotransmitters: norepinephrine and epinephrine.

There are several different types of membrane receptor proteins for each autonomic neurotransmitter (page 157).

> Responsive tissue cells possess one or more of several different types of plasma-membrane receptor proteins for these chemical messengers.
> Two types of acetylcholine receptors have been identified.
> Nicotinic receptors, found in all autonomic ganglia, respond to acetylcholine released from both sympathetic and parasympathetic preganglionic fibers.
> Muscarinic receptors, found on effector cell membranes, bind with acetylcholine released from parasympathetic postganglionic fibers.
> There are two major classes of adrenergic receptors for norepinephrine and epinephrine designated as alpha and beta receptors.

Many regions of the central nervous system are involved in the control of autonomic activities (page 158).

> Some autonomic reflexes, such as urination, defecation, and erection, are integrated at the spinal cord level, but all of these spinal reflexes are subject to control by higher levels of consciousness.
> The medulla within the brain stem is the region most directly responsible for autonomic output.
> Centers for controlling cardiovascular, respiratory and digestive activity via the autonomic system are located in the medulla.
> The hypothalamus plays an important role in integrating the autonomic, somatic and endocrine responses that automically accompany various emotional and behavioral states.
> Autonomic activity can also be influenced by the prefrontal association cortex through its involvement with emotional expression characteristic of the individual's personality.

## Somatic Nervous System (Text page 159)

Motor neurons supply skeletal muscles (page 159).

> Skeletal muscle is innervated by motor neurons, the axons of which constitute the somatic nervous system.
> The cell bodies of these motor neurons are located within the ventral horn of the spinal cord.
> The axon of a motor neuron is continuous from its origin in the spinal cord to its termination on skeletal muscle.
> Motor-neuron axon terminals release acetylcholine, which brings about excitation and contraction of the innervated muscle fibers.
> Motor neurons can only stimulate skeletal muscles.

Motor neurons are final common pathway.
The somatic system is considered to be under voluntary control, but much of the skeletal-muscle activity involving posture, balance, and stereotypical movements is subconsciously controlled.

## Acetylcholine chemically links electrical activity in motor neurons with electrical activity in skeletal muscle cells (page 159).

As an axon approaches a muscle, it divides into many terminal branches and loses its myelin sheath.
Each of these axon terminals forms a special junction, a neuromuscular junction.
A single muscle cell, a muscle fiber, is long and cylindrical in shape.
The axon terminal is enlarged into a knoblike structure, the terminal button.
The specialized portion of the muscle cell membrane immediately under the terminal button is known as the motor end plate.
Nerve and muscle cells do not actually come into direct contact at the neuromuscular junction.
A chemical messenger is used to carry the signal between the neuron terminal and the muscle fiber.
Propagation of an action potential to an axon terminal triggers the opening of voltage-gated calcium channels in the terminal button.
Opening the calcium channels permits calcium to diffuse into the terminal button, which in turn causes the release of acetylcholine from several hundred of the vesicles into the cleft.
The released ACh diffuses across the cleft and binds with specific sites.
These cholinergic receptors are of the nicotinic type.
Binding of ACh with these receptor sites induces the opening of chemical messenger-gated channels in the motor end plate.
When ACh triggers the opening of these channels, considerably more sodium moves inward than potassium outward, bringing about a depolarization of the motor end plate.
This potential change is known as the end-plate potential (EPP).
It is similar to an EPSP except that the magnitude of an EPP is much larger.
An EPP is a graded potential.
The subsequent action potential triggers contraction of the muscle.

## Acetylcholinesterase terminates acetylcholine activity at the neuromuscular junction (page 161).

The muscle's electrical response is turned off by an enzyme in the motor end-plate membrane, acetylcholinesterase (AChE) which inactivates ACh.

## The neuromuscular junction is vulnerable to several chemical agents and diseases (page 161).

The venom of black widow spiders exerts its deadly effect by causing an explosive release of ACh from storage vesicles, the most detrimental consequence of which is respiratory failure.
Botulinum toxin exerts its lethal blow by blocking the release of ACh from the terminal in the motor neuron.
Death is due to respiratory failure.
Curare reversibly binds to the ACh receptor sites on the motor end-plate.

Unlike ACh, however, curare does not alter membrane permeability nor is it inactivated by AChE.

When sufficient curare is present the person dies of respiratory paralysis.

Organophosphates irreversibly inhibit AChE.

Death from organophosphates is also due to respiratory paralysis.

In the disease myasthenia gravis, a condition characterized by extreme muscular weakness, the body erroneously produces antibodies against its own motor end-plate ACh receptors.

## KEY TERMS

Accommodation-p129
Acetylcholine-p153
Acetylcholinesterase-p161
Acuity-p121
Adrenal medulla-p157
Adrenergic fibers-p153
Alpha receptors-p158
Autonomic nervous
    system-p152
Basilar membrane-p143
Beta receptors-p158
Chemoreceptors-p120
Cholinergic fibers-p153
Cochlea-p143
Epinephrine-p157
Mechanoreceptors-p120
Motor end-plate-p159
Motor neurons-p159
Muscarinic receptors-p158

Myasthenia gravis-p163
Neuromuscular
    junction-p159
Nicotinic receptors-p158
Nociceptors-p120
Norepinephrine-p153
Opsin-p132
Optic disc-p131
Organ of Corti-p143
Osmoreceptors-p120
Ossicles-p142
Otolith organs-p148
Oval window-p143
Parasympathetic
    nervous system-p153
Parasympathetic
    tone-p156
Phasic receptors-p120
Photoreceptors-p131

Postganglionic fiber-p153
Preganglionic fiber-p153
Receptive field-p121
Receptor potential-p120
Retinene-p132
Rhodopsin-p132
Saccule-p148
Somatic nervous
    system-p153
Sympathetic nervous
    system-p153
Sympathetic tone-p156
Terminal button-p159
Terminal ganglia-p153
Thermoreceptors-p120
Tonic receptors-p120
Transduction-p118
Tympanic membrane-p142
Utricle-p148

## REVIEW EXERCISES

**True - False** (Answers on page A-11)

_____ 1. All of the nociceptors have specialized receptor structures; they are all naked nerve endings.

_____ 2. All nociceptors can be sensitized by the presence of prostaglandins.

_____ 3. Impulses from polymodal nociceptors are carried by small myelinated fibers.

_____ 4. The CNS contains a neuronal system that suppresses pain.

_____ 5. Endogenous opiates serve as analgesic neurotransmitters; they are released from a descending analgesic pathway and bind with opiate receptors on the afferent presynaptic terminal, this suppresses the release of substance Q.

_____ 6. Pain impulses originating at nociceptors are transmitted to the CNS via one of two types of efferent fibers.

_____ 7. Taste receptor cells are modified epithelial cells, with microvilli.

_____ 8. Binding of a taste-provoking chemical with a receptor cell alters its ionic channels to produce a depolarizing receptor potential.

_____ 9. Acids cause a salty taste.

_____ 10. Humans can distinguish tens of thousands of different odors.

_____ 11. Our sensitivity to a new odor rapidly diminishes after a short period of exposure to it, even though the odor source continues to be present.

_____ 12. Intense stimulation of any receptor is perceived as painful.

_____ 13. A receptor may be either a specialized ending of the afferent neuron or a separate cell closely associated with the peripheral ending of the neuron.

_____ 14. A larger receptor potential cannot bring about a larger action potential (all-or-none law), but it can induce more rapid firing of action potentials.

_____ 15. Tonic receptors are useful in situations where it is important to signal a change in stimulus intensity rather than to relay status quo information.

_____ 16. The brain cannot decode the type and location of the stimulus.

_____ 17. Each sensory neuron responds to stimulus information only within a circumscribed region of the skin surface surrounding it.

_____ 18. The pigment in the choroid and retina absorbs light after it strikes the retina to prevent reflection or scattering of light within the eye.

_____ 19. The aqueous humor is important in maintaining the spherical shape of the eyeball.

_____ 20. Iris muscles are controlled by the autonomic nervous system.

_____ 21. The strength of the lens depends on its shape, which in turn is regulated by the ciliary muscle.

_____ 22. Retinene is identical in all four photopigments, but the photoreceptor's opsin vary slightly.

_____ 23. Rhodopsin is a cone-shaped photopigment.

_____ 24. Sound waves can only travel through air.

_____ 25. Sound is characterized by its pitch (tone), intensity (loudness), and timbre (quality).

_____ 26. The tympanic membrane, collects sound waves and channels them down the external ear canal.

_____ 27. The middle ear transfers the vibratory movements of the tympanic membrane to the fluid of the inner ear.

_____ 28. The cochlea is divided throughout most of its length into four fluid-filled longitudinal compartments.

_____ 29. The fluid within the cochlear duct is called perilymph.

_____ 30. The vestibular membrane forms the floor of the cochlear duct.

_____ 31. Photoreceptors synapse with bipolar cells.

_____ 32. Rods are most abundant in the center of the retina.

_____ 33. As the axon approaches a muscle, it divides into many terminal branches and loses its myelin sheath.

_____ 34. Nerve and muscle cells occasionally come into direct contact at a neuromuscular junction.

_____ 35. A chemical messenger is used to carry the signal between the neuron terminal and the muscle fiber.

_____ 36. Propagation of an action potential to the axon terminal triggers the closing of voltage-gated calcium channels in the terminal button.

_____ 37. The released ACh diffuses across the cleft and binds with specific receptor sites.

_____ 38. The magnitude of an EPSP is much larger than an EPP.

_____ 39. The muscle's electrical response is turned on by an enzyme present in the motor end-plate membrane which inactivates ACh.

_____ 40. Motor neurons can only stimulate skeletal muscles.

_____ 41. The cells bodies of the somatic nervous system are located within the spinal cord.

_____ 42. The somatic system is considered to be under voluntary control.

_____ 43. In the autonomic nervous system there is a single neuron from the origin in the CNS to the effector organ.

_____ 44. Sympathetic nerve fibers originate in the thoracic and lumbar regions of the spinal cord.

_____ 45. Short postganglionic fibers terminate on the effector organs.

_____ 46. Sympathetic preganglionic fibers arise from the cranial and sacral areas of the CNS.

_____ 47. Parasympathetic postganglionic fibers release acetylcholine.

_____ 48. Most visceral organs are innervated by both sympathetic and parasympathetic nerve fibers.

_____ 49. Both systems increase the activity of some organs and reduce the activity of others.

_____ 50. The parasympathetic system promotes responses that prepare the body for strenuous physical activity in the face of emergency or stressful situations.

**Fill in the Blank** (Answers on page A-12)

51. Most of the eyeball is covered by a tough outer layer of connective tissue called

_____.

52. Timbre is another name for _____.

53. The ability to adjust the strength of the lens so that both near and far sources can be focused on the retina is known as _____.

54. The lens is an elastic structure consisting of transparent fibers. Occasionally these fibers become opaque so that light rays cannot pass through, a condition known as a

_____.

55. The point on the retina at which the optic nerve leaves and through which blood vessels pass is the _____.

56. A photopigment consist of an enzymatic protein called _____ combined with retinene, a derivative of vitamin A.

57. The reorganized bundles of fibers leaving the optic chiasm are known as _____.

58. Each taste bud has a small opening called the _____.

59. Sensory inputs from the chemoreceptors for taste are conveyed to the _____.

60. The axons of the olfactory receptor cells collectively form the _____.

61. The chemoreceptors for taste sensations are packaged in _____.

62. Fibers leaving the olfactory bulb travel in two different routes. (1) _____ _____, and (2) _____.

63. The olfactory mucosa contains three cell types. (1) _____, (2) _____, and (3) _____.

64. Afferent input that does reach the level of conscious awareness is known as _____ _____ and includes somatic sensation and special senses.

65. _____ receptors are rapidly adapting receptors.

66. A rise in introcular pressure is known as _____.

67. If the information detected by the receptors is propagated to the conscious level of the brain, it is called _____.

68. Pain perceived as originating in the foot by a person whose leg has been amputated at the knee is known as _____.

69. The _____ houses sensory systems for equilibrium, and provides input essential for maintenance of posture and balance.

70. The skin lining the ear canal contains modified sweat glands that produce _____, a sticky secretion that traps fine foreign particles.

71. Hair cells are specialized receptor cells that communicate via a chemical synapse with the terminals of afferent nerve fibers making up the _____.

72. Pitch discrimination depends on the shape and properties of the _____ membrane.

73. The _____ provide information about the position of the head relative to gravity and also detect changes in rate of linear motion.

74. Name the three categories of pain receptors: (1) _____, (2) _____, and (3) _____.

75. Signals arising from mechanical and thermal nociceptors are transmitted over large myelinated _____.

76. One of the neurotransmitters released from these afferent pain terminals is _____, which is believed to be unique to pain fibers.

77. All nociceptors can be sensitized by the presence of _____.

78. The built-in analgesic system is dependent on the presence of _____.

79. The _____ effect refers to a chemical or technique that brings about a desired response through the power of suggestion or distraction rather than through any direct action.

80. The skeletal muscle is innervated by _____.

81. Motor neurons are considered to be the _____ since the only way any other part of the nervous system can influence skeletal-muscle activity is by acting on these motor neurons.

82. As the axon approaches a muscle, it divides into many terminal branches and loses its myelin sheath. Each of these axon terminals forms a special junction called _____ _____.

83. A single muscle cell is called a _____.

84. The axon terminal is enlarged into a knoblike structure called a _____.

85. The specialized portion of the muscle cell membrane immediately under the terminal button is known as the _____.

86. The venom of black widow spiders exerts its deadly effect by causing an explosive release of _____ from the storage vesicles.

87. _____ exerts its lethal blow by blocking the release of ACh from the terminal button in response to an action potential in the motor neuron.

88. _____ are a group of chemicals that modify neuromuscular junction activity.

89. One disease known as _____ is a condition characterized by extreme muscular weakness.

90. The _____ innervates the effector organ. The _____ _____ synapses with the cell body of the second neuron.

91. Most sympathetic postganglionic fibers are called _____ because they release noradrenaline.

92. The terminal branches of autonomic fibers contain numerous swellings called _____.

93. About twenty percent of the adrenal medullary hormone output is norepinephrine, and the remaining eighty percent is the closely related substance _____.

94. _____ respond to acetylcholine released from both sympathetic and parasympathetic preganglionic fibers.

95. _____ are found on effector cell membranes.

**Matching:** Match the distinguishing feature to the proper division of the autonomic nervous system (Answers on page A-13)

_____ 96. Detects changes in head position away from vertical
_____ 97. Vibrates in unison with movement of stapes, to which it is attached
_____ 98. Determined by frequency
_____ 99. Contains hair cells
_____ 100. Determined by amplitude
_____ 101. Detects changes in head position away from horizontal
_____ 102. Detect angular deceleration
_____ 103. Bears the organ of Corti
_____ 104. Produces serumen
_____ 105. Equilibrates middle ear pressure with atmospheric pressure

a. oval window
b. organ of Corti
c. intensity
d. semicircular canals
e. eustachian tubes
f. ear canal
g. utricle
h. basilar membrane
i. saccule
j. pitch

Match the distinguishing feature to the proper division of the autonomic nervous system (Answers on page A-13)

_____ 106. Short cholinergic preganglionic fibers
_____ 107. Origin of preganglionic fibers is the brain and sacral region of spinal cord
_____ 108. Short cholinergic postganglionic fibers
_____ 109. Long adrenergic postganglionic fibers
_____ 110. Long cholinergic postganglionic fibers
_____ 111. Long cholinergic preganglionic fibers
_____ 112. Origin of postganglionic fibers is the terminal ganglia
_____ 113. Nicotinic receptors for neurotransmitters
_____ 114. This system dominates in emergency "fight or flight" situations
_____ 115. Muscarinic receptors for neurotransmitters

a. sympathetic system
b. parasympathetic system

**Multiple Choice** (Answers on page A-13)

_____ 116. Which receptor has a life span of about ten days?
    a. ear
    b. eye
    c. taste
    d. pain
    e. none of the above

_____ 117. Which receptors relay status quo information?
    a. photoreceptors
    b. phasic receptors
    c. nociceptors
    d. transduction receptors
    e. tonic receptors

_____ 118. Which eye disorder is characterized by the surface of the cornea being uneven?
   a. astigmatism
   b. cataract
   c. diplopia
   d. glaucoma
   e. hyperopia

_____ 119. Which part of the CNS is involved in taste?
   a. cerebellum
   b. lateral geniculate nucleus
   c. gustatory nucleus
   d. cortical gustatory area
   e. periaqueductal gray matter

_____ 120. Which of the following means nearsightedness?
   a. glaucoma
   b. astigmatism
   c. cataract
   d. myopia
   e. presbyopia

_____ 121. Which type receptor is the naked peripheral end of an afferent neuron?
   a. nociceptors
   b. mechanoreceptors
   c. photoreceptors
   d. opiate receptors
   e. vitreous receptors

_____ 122. Receptors convert various forms of energy into electrical energy. What is this conversion process called?
   a. depolarization
   b. hyperpolarization
   c. frequency modulation
   d. somesthetic propagation
   e. transduction

_____ 123. The perceptions of wetness comes from which of the following receptors?
   a. osmoreceptor
   b. chemoreceptor
   c. themroreceptor
   d. nociceptor
   e. none of the above

_____ 124. Which of the following is <u>not</u> a type of cone in the eye?
    a. blue
    b. purple
    c. green
    d. red
    e. none of the above

_____ 125. Which of the following is the area immediately surrounding the forea?
    a. blind spot
    b. optic disc
    c. optic lutea
    d. macular lutea
    e. macular disc

_____ 126. Atropine:
    a. blocks the effect of acetylcholine at muscarinic receptors
    b. blocks the effect of acetylcholine at nicotinic receptors
    c. blocks the effect of acetylcholine at beta receptors
    d. blocks the effect of norepinephrine at alpha receptors
    e. blocks the effect of norepinephrine at beta receptors

_____ 127. Curare:
    a. causes as explosive release of acetylcholine at the neuromuscular junction
    b. blocks the effects of acetylcholine at muscarinic receptors
    c. reversibly binds with acetylcholine receptor sites
    d. irreversibly inhibits acetylcholinesterase
    e. blocks the effect of norepinephrine at both beta receptors

_____ 128. Poliovirus:
    a. causes an explosive release of norepinephrine at the neuromuscular junction
    b. blocks the effect of acetylcholine at beta receptors
    c. irreversibly inhibits acetylcholinesterase
    d. destroys motor neuron cell bodies
    e. blocks the release of acetylcholine at the neuromuscular junction

_____ 129. Botulinum toxin:
    a. blocks the release of acetylcholine at the neuromuscular junction
    b. prolongs the action of acetylcholine at the neuromuscular junction
    c. blocks the effect of norepinephrine at the alpha receptors
    d. reversibly binds with acetylcholine receptor sites
    e. irreversibly inhibits Acetylcholinesterase

_____ 130. Salbutamol:
   a. activates beta adrenergic receptor sites
   b. blocks the effect of norepinephrine at beta receptors
   c. blocks the effect of acetylcholine at nicotinic receptors
   d. activates the cholinergic receptors
   e. irreversibly inhibits acetylcholinesterase

_____ 131. Mushroom poison:
   a. blocks the effects of norepinephrine at muscarinic receptors
   b. activates muscarinic receptors
   c. activates nicotinic receptors
   d. blocks the effect of norepinephrine at beta receptors
   e. causes an explosive release of norepinephrine at the neuromuscular junction

_____ 132. Black widow spider venom:
   a. blocks the effect of norepinephrine at beta receptors
   b. activates muscarinic receptors
   c. activates nicotinic receptors
   d. causes an explosive release of acetylcholine at the neuromuscular junction
   e. irreversibly inhibits acetylcholinesterase

_____ 133. Organophosphates:
   a. blocks the effect of acetylcholine at alpha receptors
   b. activates muscarinic receptors
   c. activates nicotinic receptors
   d. causes an explosive release of norepinephrine
   e. irreversibly inhibits acetylcholinesterase

_____ 134. Military nerve gas:
   a. irreversibly inhibits acetylcholinesterase
   b. causes an explosive release of acetylcholine
   c. blocks the release of acetylcholine at the neuromuscular junction
   d. reversibly binds with acetylcholine receptor sites
   e. blocks the release norepinephrine

_____ 135. Cocaine:
   a. appears to block the parasympathetic innervation of the heart
   b. appears to block the sympathetic innervation of the heart
   c. appears to inhibit acetylcholinesterase
   d. appears to  bind with muscarine
   e. appears to activate cholinergic acid

**Points to Ponder**

1. Since perception is not always reality consider these two questions:
    a. What color is a blue bird?  Be careful.
    b. As you listen to a recording of a train on your stereophonic system you hear the approach of the train in one speaker and it seems to travel across in front of you to exit in the other speaker.  How did Beethoven accomplish this phenomenon in his music?
2. Why do you think veterinarians find it difficult to diagnose toxic levels of organophosphate pesticides in animals?

**Clinical Perspectives**

1. The patient complains of severe dizziness when riding the elevator to her job on the fifteenth floor.  Explain how could she solve this problem, without walking up the stairs.
2. How does epinephrine help a person suffering from an asthma attack?
3. How does epinephrine help a person who has just been stung by a wasp and is hypersensitive to bee stings?
4. A patient is accidentally given curare.  How would you keep this person alive?  Would the same treatment work for an accidental dose of organophosphates?

**Experiment of the Day**

Place your index finger against the corner of your closed eye.  Apply short bursts of gentle pressure on the eyeball.  Describe and explain this phenomenon.

# MUSCLE PHYSIOLOGY

**6**

## CHAPTER OVERVIEW

Muscle cells, subsequently muscle fibers, are specialized to contract. Again the special-ization is the extensive development of a basic fundamental function of most cells. Movement of the body, manipulation of objects and movement of the contents of the body are all important in maintaining homeostasis.

Muscles are classified by their appearance, their function, and the nerves that innervate them. The contractile elements are actin and myosin. Individual muscle fibers either contract or they do not contract.

Muscles are under the control and coordination of the CNS. These contraction specialists respond to an electrical signal and convert chemical energy into mechanical energy.

Again, we are looking at a body system working to maintain homeostasis which is essential for the survival of all the cells. Considering all three types, muscles play a very large and important role in maintaining homeostasis.

## CHAPTER OUTLINE

### Introduction (Text page 169)

Although there are three muscle types that are structurally and functionally distinct, they can be classified in two different ways according to their common characteristics.

First, muscles are categorized as striated (skeletal and cardiac muscle) or unstriated (smooth).

Second, muscles are categorized as voluntary (skeletal muscle) or involuntary (cardiac and smooth muscle).

Voluntary muscles are innervated by the somatic nervous system.

Involuntary muscles are innervated by the autonomic nervous system.

## Structure of Skeletal Muscle (Text page 169)

Skeletal-muscle fibers have a highly organized internal arrangement that creates a striated appearance (page 169).

A single skeletal-muscle cell is known as a muscle fiber.

A skeletal muscle consists of a number of muscle fibers lying parallel to each other.

The most predominant structural feature of a skeletal-muscle fiber is the presence of numerous myofibrils.

Each myofibril consists of a regular arrangement of the proteins myosin (thick filaments) and actin (thin filaments).

The bands of all the myofibrils lined up parallel to each other collectively lead to the striated appearance of a skeletal-muscle fiber.

An A band consists of a stacked set of thick filaments along with the portions of the thin filaments that overlap on both ends of the thick filaments.

The thick filaments are found only in the A band and extend its entire width.

The lighter area within the middle of the A band, where the thin filaments do not reach, is known as the H zone. Only the central portions of the thick filaments are found in this region.

The I band consists of the remaining portion of the thin filaments that do not project into the A band.

The I band contains only thin filaments but not the entire length of these filaments.

In the middle of each I band is a dense, vertical Z line.

The area between two Z lines is called a sarcomere, which is the functional unit of skeletal muscle.

Myosin forms the thick filaments, whereas actin is the main structural component of the thin filaments (page 171).

Each thick filaments is composed of several hundred myosin molecules packed together in a specific arrangement.

A myosin molecule is a protein consisting of two identical subunits shaped somewhat like a golf club.

The globular heads form the cross bridges between thick and thin filaments.

Each cross bridge has two important sites crucial to the contractile process: an actin binding site and a myosin ATPase site.

Thin filaments are composed of three proteins--actin, tropomyosin, and troponin.

Actin molecules are spherical in shape.

The backbone of a thin filament is formed by actin molecules joined into two strands and twisted together.

Tropomyosin molecules are threadlike proteins that lie end to end alongside the grove of the actin spiral.

Tropomyosin covers the actin sites that bind with the cross bridges, thus blocking the interaction that leads to muscle contraction.

Tropomyosin is stabilized in this blocking position by troponin molecules, which fasten down the ends of each tropomyosin molecule.

When calcium binds to troponin, the shape of this protein is changed in such a way that tropomyosin is allowed to slide away from its blocking position.

With tropomyosin out of the way, actin and myosin can bind and interact at the cross bridges resulting in muscle contraction.

# Molecular Basis of Skeletal-Muscle Contraction (Text page 172)

Cycles of cross-bridge binding and bending pull the thin filaments closer together between the thick filaments during contraction (page 172).

> The thin filaments on each side of a sarcomere slide inward toward the A band's center, during contraction.
>
> As they slide inward, the thin filaments pull the Z lines to which they are attached closer together, so the sarcomere shortens.
>
> This is known as the sliding-filament mechanism of muscle contraction.
>
> When myosin and actin make contact at a cross bridge, the conformation of the bridge is altered so that it bends inward.
>
> This so-called power stroke of a cross bridge pulls the thin filament to which it is attached inward.
>
> Complete shortening is accomplished by repeated cycles of cross-bridge binding and bending.

## Calcium is the link between excitation and contraction (page 175).

> Skeletal muscles are stimulated to contract by release of acetylcholine at neuromuscular junctions.
>
> At each junction of an A band and an I band, the surface membrane dips into the muscle fiber to form a transverse tubule (T tubule), which runs perpendicularly from the surface of the muscle cell membrane into the central portions of the muscle fiber.
>
> The T tubule provides a means of rapidly transmitting the surface electrical activity into the central portions of the fiber.
>
> The sarcoplasmic reticulum is a modified endoplasmic reticulum that consists of a fine network of interconnected tubules surrounding each myofibril.
>
> Separate segments of sarcoplasmic reticulum are wrapped around each A band and I band.
>
> The ends of each segment expand to form saclike regions, the lateral sacs, which lie in close proximity to the adjacent T tubules.
>
> The sarcoplasmic reticulum lateral sacs store calcium.
>
> Spread of an action potential down a T tubule triggers the release of calcium from the sarcoplasmic reticulum into the cytosol.
>
> This released calcium, by slightly repositioning the troponin and tropomyosin molecules, exposes the binding sites on the actin molecules so that they can link with the myosin cross bridges at their complementary binding sites.
>
> The ATPase site is an enzymatic site that can bind the energy carrier ATP, and split it into ADP and inorganic phosphate, yielding energy in the process.
>
> In skeletal muscle, magnesium must be attached to ATP before myosin ATPase can split the ATP.
>
> The generated energy is stored within the cross bridge to produce a high-energy form of myosin.
>
> When the muscle fiber is excited, calcium pulls the troponin-tropomyosin complex out of its blocking position so that the energized myosin cross bridge can bind with an actin molecule.
>
> The energy stored within the myosin cross bridge is released to cause the cross bridge bending responsible for the power stroke that pulls the thin actin filament inward.
>
> The contractile process is turned off when the calcium is returned to the lateral sacs upon cessation of local electrical activity.

When acetylcholinesterase removes ACh from the neuromuscular junction, the muscle-fiber action potential ceases.

Removal of cytosolic calcium allows the troponin-tropomyosin complex to slip back into its blocking position.

Relaxation has occurred.

### Contractile activity far outlasts the electrical activity that initiated it (page 178).

The time delay between stimulation and the onset of contraction is known as the latent period.

The time from the onset of contraction until peak tension is developed--the contraction time.

The time from peak tension until relaxation is complete is called the relaxation time.

## Skeletal Muscle Mechanics (Text page 179)

### Whole muscles are groups of muscle fibers bundled together by connective tissue and attached to bones by tendons (page 179).

Groups of muscle fibers are organized into whole muscles.

Each muscle is covered by a sheath of connective tissue that penetrates from the surface into the muscle to envelop each individual fiber and divides the muscle into columns or bundles.

The connective tissue further extends beyond the ends of the muscle to form tough, collagenous tendons that attach the muscle to bones.

### Contractions of a whole muscle can be of varying strength (page 179).

Muscle fibers are arranged into whole muscles where they can function cooperatively to produce contractions of variable grades of strength stronger than a twitch.

Two primary factors can be adjusted to accomplish gradation of whole-muscle tension: (1) the number of muscle fibers contracting within a muscle and (2) the tension produced by each contracting fiber.

### The number of fibers contracting within a muscle depends on the extent of motor unit recruitment (page 179).

Larger muscles consisting of more muscle fibers are obviously capable of generating more tension than are smaller muscles with fewer fibers.

One motor neuron innervates a number of muscle fibers, but each muscle fiber is supplied by only one motor neuron.

This functional unit--one motor neuron plus all of the muscle fibers it innervates--is called a motor unit.

For a weak contraction of the whole muscle, only one or a few of its motor units are activated.

For stronger and stronger contractions, more and more motor units are recruited, or stimulated to contract, a phenomenon known as motor-unit recruitment.

To delay or prevent fatigue during a sustained contraction involving only a portion of a muscle's motor-units, as is necessary in muscles supporting the weight of the body against the force of gravity, asynchronous recruitment of motor units takes place.

The body alternates motor-unit activity to give motor units that have been active an opportunity to rest while others take over.

The frequency of stimulation can influence the tension developed by each muscle fiber (page 180).

Various factors influence the extent to which tension can be developed including: (1) the frequency of stimulation, (2) the length of the fiber at the onset of contraction, (3) the extent of fatigue, and (4) the thickness of the fiber.

Even though a single action potential in a muscle fiber produces only a twitch, greater tension can be achieved by repetitive stimulation of the fiber.

If the muscle fiber is stimulated a second time before it has completely relaxed from the first twitch, a second action potential occurs that causes a second contractile response. The two twitches resulting from the two action potentials add together, or sum, to produce greater tension than produced by a single action potential.

This is known as twitch summation.

There is an optimal muscle length at which maximal tension can be developed upon a subsequent contraction (page 182).

For every muscle there is an optimal length at which maximal force can be achieved upon a subsequent tetanic contraction.

In the body the muscles are so positioned that their relaxed length is approximately their optimal length.

The two primary types of contraction are isotonic and isometric (page 182).

The end of the muscle attached to the more stationary part of the skeletal is called the origin, whereas the end attached to the skeletal part that moves is referred to as the insertion.

In an isotonic contraction, muscle tension remains constant as the muscle changes length.

In an isometric contraction, the muscle is prevented from shortening, so the development of tension occurs at constant muscle length.

With concentric contractions the muscle shortens, whereas with eccentric contractions the muscle lengthens because it is being stretched by an external force while contracting.

The velocity of shortening is related to the load (page 184).

The greater the lead, the lower the velocity at which the muscle fiber shortens during an isotonic tetanic contraction.

## Skeletal Muscle Metabolism and Fiber Types (Text page 185)

Muscle fibers have alternate pathways for forming ATP (page 185).

Three different steps in the contraction-relaxation process require ATP: (1) the splitting of ATP by myosin ATPase; (2) the binding of a fresh molecule of ATP to myosin; and (3) the active transport of calcium back into the sarcoplasmic reticulum during relaxation depends on energy derived from the breakdown of ATP.

Only limited stores of ATP are immediately available in muscle, but three pathways supply additional ATP as needed during muscle contraction: (1) transfer of a high-energy phosphate from creatine phosphate to ADP; (2) oxidative phosphorylation; (3) glycolysis.

Creatine phosphate is the first energy storehouse tapped at the onset of contractile activity.

The energy released from the hydrolysis of creatine phosphate, along with the phosphate, can be donated directly to ADP to form ATP.

Oxidation phosphorylation takes place within the muscle mitochondria if sufficient oxygen is present.

During light exercise to moderate exercise, muscle cells are able to form sufficient amounts of ATP through oxidative phosphorylation to keep pace with the modest energy demands of the contractile machinery for prolonged periods of time.

Activity that can be supported in this way is known as endurance type exercise or aerobic exercise.

The oxygen required for oxidative phosphorylation is primarily delivered by the blood.

When oxygen delivery or oxidative phosphorylation cannot keep up with the demand for ATP formation as the intensity of exercise increases, the muscle fibers rely increasingly on glycolysis to generate ATP.

Glycolysis alone has two advantages: (1) glycolysis can form ATP in the absence of oxygen, and (2) it can proceed more rapidly than oxidative phosphorylation because it requires fewer steps.

Even though anaerobic glycolysis provides a means of performing intense exercise, using this pathway has two consequences: (1) large amounts of nutrient fuel must be processed, and (2) the end product of anaerobic glycolysis, pyruvic acid, is converted to lactic acid when it cannot be further processed by the oxidative phosphorylation pathway.

High-intensity or anaerobic exercise can be sustained for only a short duration in contrast to the body's prolonged ability to sustain aerobic activities.

## Increased oxygen consumption is necessary to recover from exercise (page 185).

The best known reason of elevated oxygen uptake during recovery is repayment of an oxygen debt that was incurred during exercise.

Part of the extra oxygen uptake during recovery is not directly related to the repayment of energy stores but instead is the result of a general metabolic disturbance following exercise.

## There are three types of skeletal muscle fibers based on differences in ATP hydrolysis and synthesis (page 186).

Based on their biochemical capacities, there are three major types of muscle fibers: (1) slow-oxidative (type I) fibers; (2) fast-oxidative (type IIa) fibers, and (3) fast-glycolytic (type IIb) fibers.

Fast fibers have higher myosin ATPase activity than slow fibers.

Fibers also differ in their ATP-synthesizing ability.

Oxidative types of muscle fibers are more resistant to fatigue than are glycolytic fibers.

Oxidative fibers also have a high myoglobin content and are referred to as red fibers.

The glycolytic fibers contain very little myoglobin and therefore are pale in color, so they are sometimes called white fibers.

In humans, most of the muscles contain a mixture of all three fiber types.

## Muscle fibers adapt considerably in response to the demands placed on them (page 187).

Regular endurance exercise induces metabolic changes within the oxidative fibers, which are the ones primarily recruited during aerobic exercises.

Muscles so adapted are better able to endure prolonged activity without fatiguing, but they do not change in size.

The actual size of muscles can be increased by regular bouts of anaerobic, short-duration, high-intensity resistance training, such as weight lifting.

The resulting muscle enlargement comes primarily from an increase in diameter of the fast-glycolytic fibers that are called into play during such powerful contractions.

The resultant bulging muscles are better adapted to activities that require intense strength for brief periods, but endurance has not been improved.

# Control of Motor Movement (Text page 187)

### Many inputs influence motor unit output (page 187).

Three levels of input control motor neuron output: (1) input from afferent neurons, usually through intervening interneurons, at the level of the spinal cord--that is, spinal reflexes; (2) input from the primary motor cortex, the corticospinal motor system; and (3) input from the multineuronal motor system.

The corticospinal system primarily mediates performances of fine, discrete, voluntary movements of the hands and fingers.

The multineuronal system is primarily concerned with regulation of overall body posture involving involuntary movements of large muscle groups of the trunks and limbs.

### Muscle spindles and the Golgi tendon organs provide afferent information essential for controlling skeletal muscle activity (page 189).

The CNS must know the starting position of your body to appropriately program muscle activity.

The CNS must be constantly apprised of the progression of movement it has initiated so that it can make adjustments as needed.

Your brain receives this information, which is known as proprioceptive input, from receptors in your eyes, joints, vestibular apparatus, and skin, as well as from the muscles themselves.

Muscle length is monitored by muscle spindles, whereas changes in muscle tension are detected by Golgi tendon organs.

Muscle spindles consist of collections of specialized muscle fibers known as intrafusal fibers which lie within spindle-shaped connective-tissue capsules parallel to the "ordinary" extrafusal fibers.

The efferent neuron that innervates a muscle spindle's intrafusal fibers is known as a gamma motor neuron, whereas the motor neurons that supply the ordinary extrafusal fibers are designated as alpha motor neurons.

The afferent sensory endings are wrapped around the central portions of the intrafusal fibers; they detect changes in the length of the fibers during stretching as well as the speed with which it occurs.

This stretch reflex serves as a local negative-feedback mechanism to resist any passive changes in muscle length so that optimal resting length can be maintained.

The primary purpose of the stretch reflex is to resist the tendency for the passive stretch of extensor muscles caused by gravitational forces when a person's standing upright.

Golgi tendon organs are located in the tendons of the muscle, where they are able to respond to changes in the muscle's externally applied tension rather than to changes in its length.

It is essential that motor control systems be apprised of the tension actually achieved so that adjustments can be made if necessary.

The Golgi tendon organs consist of endings of afferent fibers entwined within bundles of connective-tissue fibers that make up the tendon.

When the extrafusal muscle fibers contract, the entwined Golgi organ afferent receptor endings are stretched, causing the afferent fibers to fire; the frequency of firing is directly related to the tension developed.

Input from the activated Golgi tendon organs counterbalances excitatory inputs to the alpha motor neurons.

This inhibitory response halts further contraction and brings about sudden reflex relaxation, thus helping prevent damage to muscle or tendons from excessive, tension-developing muscle contractions.

# Smooth and Cardiac Muscle (Text page 192)

## Smooth and cardiac muscle share some basic properties with skeletal muscle (page 192).

Smooth muscle and cardiac muscle share some basic properties with skeletal muscle, but each also displays unique characteristics.

## Smooth muscle cells are small and unstriated (page 194).

The majority of smooth muscle cells are found in the walls of hollow organs and tubes.

Smooth muscle cells are spindle-shaped, have a single nucleus, and are considerably smaller than skeletal muscle cells.

Also, a single smooth muscle cell does not extend the full length of a muscle.

Three types of filaments are found in a smooth muscle cell: (1) thick myosin filaments; (2) thin actin-filaments; and (3) unique filaments of intermediate size.

## Smooth muscle cells are turned on by calcium-dependent phosphorylation of myosin (page 195).

Smooth muscle myosin is able to interact with actin only when the myosin is phosphorylated.

Phosphorylated myosin then binds with actin so that crossbridge cycling can begin.

When calcium is removed the muscle relaxes.

## Most groups of smooth muscle tissue are capable of self-excitation (page 195).

Smooth muscle is grouped into two categories--multiunit and single-unit smooth muscle based on differences in how the muscle fibers become excited.

Skeletal muscle and multiunit smooth muscle are both neurogenic; that is, they depend on their nerve supply to initiate contraction.

Multiunit smooth muscle is supplied by the involuntary autonomic nervous system.

Multiunit smooth muscle is found (1) in the walls of large blood vessels; (2) in large airways to the lungs; (3) in the muscle of the eye that adjusts the lens; (4) in the iris of the eye; and (5) at the base of hair follicles.

Most smooth muscle is of the single-unit variety.

Single-unit is alternately called visceral smooth muscle because it is found in the walls of the hollow organs or viscera (for example, the digestive, reproductive, and urinary tracts and small blood vessels).

The muscle fibers in single-unit smooth muscle are electrically linked by gap junctions.

When an action potential occurs anywhere within a sheet of single-unit smooth muscle, it is quickly propagated throughout the entire group of interconnected cells, which then contract as a single coordinated unit, known as a functional syncytium.

Single-unit smooth muscle is self-excitable rather than requiring nervous stimulation for contraction.

There are two major types of spontaneous depolarizations displayed by self-excitable cells: pacemaker activity and slow-wave potentials.

In pacemaker activity the membrane potential gradually depolarizes on its own.

When the membrane has depolarized to threshold, an action potential is initiated.

Self-generated action potentials are cyclically produced.

Slow-wave potentials are gradually alternating hyperpolarizing and depolarizing swings in the potential caused by automatic cyclical changes in the rate at which sodium ions are actively transported across the membrane.

If threshold is reached, action potentials occurs at the peak of a depolarizing swing.

Such nerve-independent contractile activity initiated by the muscle itself is called myogenic activity.

## Gradation of single-unit smooth muscle contraction differs considerably from that of skeletal muscle (page 196).

In smooth muscle, the gap junctions ensure than an entire smooth muscle mass contracts as a single unit, making it impossible to vary the number of muscle fibers contracting.

The portion of cross bridges activated and the tension subsequently developed in single-unit smooth muscle can be graded by varying the cytosolic calcium concentration.

Smooth muscle is typically innervated by both branches of the autonomic nervous system.

In single-unit smooth muscle, this nerve supply does not initiate contraction, but it can modify the rate and strength of contraction, either enhancing or retarding the inherent contractile activity of a given organ.

The ability of a considerably stretched smooth muscle fiber to still develop tension is important, because the smooth muscle fibers within the wall of a hollow organ are progressively stretched as the volume of the organ's contents increases.

## Smooth muscle is slow and economical (page 198).

A smooth muscle contractile response proceeds at a more leisurely pace than does a skeletal muscle twitch.

Because of the low rate of cross bridge cycling, cross bridges are maintained in the attached state for a longer period of time during each cycle; that is, the cross bridges "latch onto" the thin filaments for a longer time each cycle.

This so-called latch phenomenon enables smooth muscle to maintain tension with comparatively less ATP consumption.

## Cardiac muscle blends features of both skeletal and smooth muscle (page 198).

Cardiac muscle, found only in the heart, structurally and functionally shares characteristics with both skeletal and single-unit smooth muscle.

Cardiac muscle is striated.

Cardiac thin filaments contain troponin and tropomyosin.

Like single-unit smooth muscle, the heart displays pacemaker activity.

Cardiac cells are interconnected by gap junctions.

The heart is innervated by the autonomic nervous system, which along with certain hormones and local factors, can modify the rate and strength of contraction.

Unique to cardiac muscle, the cardiac fibers are joined together in a branching network.

## KEY TERMS

Actin-p172
Contraction time-p178
Cross bridges-p171
Extrafusal fibers-p190
Intrafusal fibers-p190
Latent period-p178
Motor units-p179

Muscle fibers-p169
Muscle spindle-p190
Myofibrils-p169
Myogenic activity-p196
Myoglobin-p186
Myosin-p171
Oxygen debt-p186

Power stroke-p173
Relaxation time-p178
Sarcomere-p171
Transverse tubule-p175
Tropomyosin-p172
Troponin-p172
Twitch summation-p180

## REVIEW EXERCISES

**True - False** (Answers on page A-14)

_____ 1. The corticospinal system primarily mediates performance of fine, discrete, voluntary movements of the hands and fingers.

_____ 2. Muscle spindles are specialized muscle fibers known as extrafusal fibers.

_____ 3. The efferent neuron that innervates a muscle spindle's intrafusal fibers is known as a gamma motor neuron.

_____ 4. The stretch reflex serves as a local positive-feedback mechanism to resist any passive changes in muscle length.

_____ 5. Golgi tendon organs respond to changes in muscle length.

_____ 6. Hemiplegia is the paralysis of one side of the body.

_____ 7. Ordinary muscle fibers are known as extrafusal fibers.

_____ 8. Smooth-muscle cells are spindle-shaped with multiple nuclei.

_____ 9. Contractile activity initiated by the muscle itself is called myogenic activity.

_____ 10. Cardiac cells are interconnected by gap junctions.

_____ 11. The portion of cross bridges activated and the tension subsequently developed in a single-unit smooth muscle can be graded by varying the cytosolic magnesium concentration.

_____ 12. The splitting of ATP by actin ATPase provides the energy for the power stroke of the cross bridge.

_____ 13. Creatine phosphate is the first energy storehouse tapped at the onset of contractile activity.

_____ 14. Hemoglobin increases the rate of oxygen transfer from the blood into the muscle fiber.

_____ 15. The end product of anaerobic glycolysis is lactic acid.

_____ 16. Psychological fatigue occurs when a peron stops exercising even though the muscles are still able to perform.

_____ 17. Glycolytic fibers are more resistant to fatigue.

_____ 18. A single skeletal-muscle cell is known as a myofibril.

_____ 19. Thick filaments are special assemblies of the protein myosin.

_____ 20. Each myofibril consists of a regular arrangement of the proteins myosin and actin.

_____ 21. Tropomyosin molecules lie end to end alongside the groove of the myosin spiral.

_____ 22. The globular heads of the myosin molecule form the cross bridges between the thick and thin filaments.

_____ 23. The area between two Z lines is called a sarcomere.

_____ 24. When magnesium binds to troponin, the shape of this protein is changed in a way that tropomyosin is allowed to slide away from its blocking position.

_____ 25. One motor neuron plus all the muscle fibers it innervates is called a motor unit.

_____ 26. A single action potential in a muscle fiber produces a brief, weak contraction known as a twitch.

_____ 27. The epithelial tissue further extends beyond the ends of the muscle to form a tough, collagenous tendon.

_____ 28. The asynchronous recruitment of motor units produces muscle spasms.

_____ 29. Thickness of the fiber influences the extent to which tension can be developed.

_____ 30. The phenomenon known as summation is when more and more motor units are stimulated for stronger contractions.

_____ 31. A sustained contraction of maximal strength is known as tetanus.

_____ 32. The end of the muscle attached to the more stationary part of the skeleton is known as the insertion.

_____ 33. In eccentric contractions the muscle shortens.

_____ 34. At each junction of an A band and an I band, the surface membrane dips into the muscle fiber to form a transverse tubule.

_____ 35. The sarcoplasmic tubule is a modified endoplasmic reticulum.

_____ 36. T tubules store calcium.

_____ 37. In skeletal muscle, magnesium must be attached to ATP before myosin ATPase can split the ATP.

_____ 38. The time delay between stimulation and the onset of contraction is known as the contraction time.

_____ 39. The contractile process is turned off when calcium is returned to the latent sacs upon cessation of local electrical activity.

_____ 40. Removal of cytosolic potassium allows the troponin-tropomyosin complex to slip back into its blocking position.

**Fill in the Blank** (Answers on page A-15)

41. _____ refers to the series of events linking muscle excitation to muscle contraction.

42. At each junction of an A band and an I band, the surface membrane dips into the muscle fiber to form a _____, which runs perpendicularly from the surface of the muscle cell membrane into the central portions of the muscle fiber.

43. The _____ is a modified endoplasmic reticulum that consist of a fine network of interconnected tubules surrounding each myofibril.
44. The ends of each sarcoplasmic reticulum segments expand to form saclike regions, the _____, which lie in close proximity to the adjacent T tubules.
45. A myosin cross bridge has two special sites, an _____ site and an _____ site.
46. The necessity for ATP in the separation of myosin and actin is amply demonstrated by the phenomenon of _____.
47. The time delay between stimulation and the onset of contraction is known as the _____.
48. _____ exhibits properties partway between skeletal muscle and single unit smooth muscle.
49. _____ means they depend on their nerve supply to initiate contraction. Examples are skeletal muscle and multiunit smooth muscle.
50. A group of interconnected muscle cells that function electrically and mechanically as a unit is known as a _____.
51. _____ are gradually alternating hyperpolarizing and depolarizing swings in potential caused by automatic cyclical changes in the rate at which sodium ions are actively transported across the membrane.
52. _____ enables smooth muscle to maintain tension with comparatively less ATP consumption, because each cross-bridge cycle uses up one molecule of ATP.
53. A single skeletal-muscle cell is known as _____.
54. The most predominant structural feature of a skeletal-muscle fiber is the presence of numerous _____.
55. An _____ consists of a stacked set of thick filaments along with the portions of the thin filaments that overlap on both ends of the thick filaments.
56. The area between two Z lines is called a _____.
57. A _____ molecule is a protein consisting of two identical subunits, each shaped somewhat like a golf club. _____ molecules are spherical in shape.
58. _____ molecules are threadlike proteins that lie end to end alongside the groove of the actin spiral.
59. _____ and _____ are often referred to as regulatory proteins.
60. The efferent neuron that innervates a muscle spindle's intrafusal fibers is known as a _____, whereas the motor neurons that supply the ordinary extrafusal fibers are designated as _____.
61. _____ are located in the tendons of the muscle, where they are able to respond to changes in the muscle's externally applied tension rather than to changes in its length.
62. One motor neuron plus all of the muscle fibers it innervates is called a _____.
63. Cross-bridge activity produces _____ within the sarcomeres.

64. The noncontractile tissues are referred to as the _____ component of the muscle.

65. The end of the muscle attached to the more stationary part of the skeleton is called the _____, whereas the end attached to the skeletal part that moves is referred to as the _____.

66. In an _____ contraction, muscle tension remains constant as the muscle changes length.

67. There are actually two types of isotonic contractions, _____ and _____.

68. _____ is the first energy storehouse tapped at the onset of contractile activity.

69. _____ takes place within the muscle mitochondria if sufficient oxygen is present.

70. _____ increases the rate of oxygen transfer from the blood into muscle fibers.

71. _____ atrophy occurs when a muscle is not used for a long period of time even though the nerve supply is intact.

72. _____ is a hereditary pathological condition characterized by progressive degeneration of contractile elements, which are ultimately replaced by fibrous tissue.

**Matching:**  Match the characteristic to the type of fiber (Answers on page A-15)

_____ 73. Intermediate intensity of contraction

_____ 74. High intensity of contraction

_____ 75. High resistance to fatigue

_____ 76. Low resistance to fatigue

_____ 77. Low intensity of contraction

_____ 78. Low oxidative phosphorylation capacity

_____ 79. Slow speed of contraction

_____ 80. White fibers

_____ 81. Intermediate amount of enzymes for anaerobic glycolysis

_____ 82. Low myoglobin content

_____ 83. Low myosin-ATPase activity

a, Fast-Glycolytic Type IIb
b. Fast-Oxidative Type IIa
c. Slow-Oxidative Type I

**Multiple Choice** (Answers on page A-15)

_____ 84. Which of the following consists of a stacked set of thick filaments along with portions of the thin filaments that overlap on both ends of the thick filaments?
   a. Z line
   b. M line
   c. A band
   d. I band
   e. H zone

_____ 85. Which of the following are threadlike proteins that lie end to end alongside the groove of the actin spiral?
- a. tropomyosin
- b. troponin
- c. myosin
- d. tubulin
- e. myoglobin

_____ 86. Which of the following ions must be attached to ATP before myosin ATPase can split the ATP in skeletal muscle?
- a. magnesium
- b. calcium
- c. potassium
- d. sodium
- e. calmodulin

_____ 87. Which of the following represents the region where only the central portions of the thick filaments are found?
- a. Z line
- b. M line
- c. A band
- d. I band
- e. H zone

_____ 88. Which of the following pairs of cross-bridge sites is crucial to the contractile process?
- a. troponin binding site and myosin ATPase site.
- b. tropomyosin binding site and actin ATPase site
- c. myosin binding site and myosin ATPase site
- d. actin binding site and myosin ATPase site
- e. myoglobin binding site and troponin ATPase site

_____ 89. Which of the following binds with troponin in such a way that tropomyosin is allowed to move from its blocking position?
- a. magnesium
- b. calcium
- c. potassium
- d. sodium
- e. troponin

_____ 90. Which of the following types of muscle has no gap junctions?
- a. multiunit smooth
- b. skeletal
- c. single-unit smooth
- d. cardiac
- e. all of the above

_____ 91. Which of the following types of muscle has cross bridges turned on by calcium?
    a. multiunit smooth
    b. skeletal
    c. single-unit smooth
    d. cardiac
    e. all of the above

_____ 92. Which of the following is innervated by the somatic nervous system?
    a. multiunit smooth
    b. skeletal
    c. single-unit smooth
    d. cardiac
    e. all of the above

_____ 93. Which of the following is the functional unit of a skeletal muscle?
    a. Golgi organ
    b. thick filaments
    c. sarcomere
    d. sliding filaments
    e. transverse-contractile coupler

**Points to Ponder**
1. Can you think of an example where involuntary muscles are stimulated to contract voluntarily?
2. How can an average sized woman beat a contestant for Mr. Universe in arm wrestling?
3. How does weight-lifting shape up as a healthy form of exercise?
4. Why do you think the average life expectancy for NFL players is only 57 years?

**Clinical Perspectives**
1. Explain why heat and/or cold packs are used to treat the so-called "pulled muscle."

**Experiment of the Day**
Go to the meat department of your favorite grocery store and explain the appearance of the meats in terms of muscle structure.

# CARDIAC PHYSIOLOGY

7

## CHAPTER OVERVIEW

The importance of the circulatory system, particularly the heart, in maintaining homeostasis has been known to some degree by most of us since our childhood days. This heart-homeostasis relationship is probably the best known aspect of homeostasis. From the bending and twisting of an embryonic blood vessel an organ develops that will pump the blood. Complete with muscular walls and valves, the heart responds to the autorhythmicity of the sinoatrial node. Using specialized conductive cells, action potentials are spread throughout the organ and coordinated contractions occur. After a brief period of relaxation the cycle begins again.

The electrical and mechanical activities of the heart produce sounds, pressures, and electrical changes that can be detected on the surface of the body. Electrocardiograph is an important tool in monitoring cardiac activities.

To meet the changing homeostasis needs of all the cells, the heart must be able to change its output. Cardiac output is controlled through changes in heart rate and stroke volume. While the heart is working for all the cells it must also nourish itself. This is one of the most frequent areas of complications in the circulatory system.

Even though the heart is emphasized in science and culture, in reality it is but a part of an organ system controlled by the nervous and endocrine systems. The remarkable properties of the heart are specializations of the general properties and functions of all cells. The circulatory system aids in maintaining homeostasis which is essential for the survival of cells.

# CHAPTER OUTLINE

## Introduction (Text page 204)

The circulatory system consists of the heart, the blood vessels and the blood.

The pulmonary circulation of a closed loop of vessels carrying blood between the heart and lungs, whereas the systemic circulation consists of a circuit of vessels carrying blood between the heart and organ systems.

## Anatomical Considerations (Text page 205)

### The heart is located in the middle of the chest cavity (page 205).

The heart is situated at an angle under the sternum so that its base lies predominately to the right and the apex to the left of the sternum.

The fact that the heart is positioned between two bony structures, the sternum and vertebrae, make it possible to manually drive blood out of the heart.

### The heart is a dual pump (page 206).

Even though anatomically the heart is a single organ, the right and left sides of the heart function as two separate pumps.

The upper chambers, the atria, receive blood returning to the heart and transfer it to the lower chambers, the ventricles, which pump the blood from the heart.

The vessels that return the blood from the tissues to the atria are veins, and those that carry blood away from the ventricles to the tissues are arteries.

The right side of the heart pumps blood into the pulmonary circulation.

The right ventricle pumps blood out through the pulmonary artery to the lungs.

The left side of the heart pumps blood into the systemic circulation.

The large artery carrying blood away from the left ventricle is the aorta.

Both sides of the heart simultaneously pump equal amounts of blood.

The volume of low-oxygenated blood being pumped to the lungs by the right side of the heart soon becomes the same volume of high-oxygenated blood being delivered to the tissues by the left side of the heart.

The pulmonary circulation is a low-pressure, low resistance system, whereas the systemic circulation is a high-pressure, high-resistance system.

### Heart valves ensure that the blood flows in the proper direction through the heart (page 207).

Blood flows through the heart in one fixed direction from veins to atria to ventricles to arteries.

Two of the heart valves, the right and left atrioventricular (AV) valves, are positioned between the atrium and the ventricle on the right and left sides, respectively.

The two remaining heart valves, the aortic and pulmonary valves, are located at the junction where the major arteries leave the ventricles.

They are known as semilunar valves because they are composed of three cusps, each resembling a shallow half-moon shaped pocket.

The heart walls are composed primarily of spirally arranged cardiac muscle fibers interconnected by intercalated discs (page 209).

The heart wall consists of three distinct layers: (1) The endocardium is a thin inner layer of endothelium that lines the entire circulatory system. (2) The myocardium, the middle layer composed of cardiac muscle, constitutes the bulk of the heart wall. (3) The epicardium is a thin external membrane covering the heart.

The myocardium consists of interlacing bundles of cardiac muscle fibers arranged spirally around the circumference of the heart.

When the ventricular muscle contracts and shortens, the diameter of the ventricular chambers is reduced while the apex is simultaneously pulled upward toward the top of the heart in a rotating manner.

The individual cardiac muscle cells are interconnected to form branching fibers, with adjacent cells joined end to end at specialized structures known as intercalated discs. Within an intercalated disc, there are two types of membrane junctions: desmosomes and gap junctions.

Cardiac muscle is capable of generating action potentials without nervous stimulation. There are no gap junctions between atrial and ventricular contractile cells, and furthermore, these muscle masses are separated by the electrically nonconductive fibrous skeleton that surrounds the valves.

An important specialized conducting system is present to facilitate and coordinate the transmission of electrical excitation from the atria to the ventricles.

Either all of the cardiac muscle fibers contract or none of them do.

Gradation of cardiac contraction is accomplished by varying the strength of contraction of all the cardiac muscle cells.

The heart is enclosed by the pericardial sac (page 211).

The heart is enclosed in the double-walled, membranous pericardial sac.

The sac is lined by a membrane that secretes a thin pericardial fluid, which provides lubrication to prevent friction between the pericardial layers as they glide over each other.

# Electrical Activity of the Heart (Text page 211)

The sinoatrial node is the normal pacemaker of the heart (page 211).

The heart contracts or beats rhythmically as a result of action potentials that it generates by itself, a property known as autorhythmicity.

There are two types of cardiac muscle cells: (1) Ninety-nine percent of the cardiac muscle cells are contractile cells. (2) The autorhythmic cells do not contract, but instead are specialized for initiating and conducting the action potentials responsible for contraction.

The cardiac autorhythmic cells display pacemaker activity.

The cardiac cells capable of autorhythmicity are found in the following specific locations: (1) the sinoatrial node, (2) the atrioventricular node, (3) the bundle of His, and (4) the Purkinje fibers.

Various autorhythmic cells differ in the rates at which they are normally capable of generating action potentials.

The SA node, which normally exhibits the fastest rate of autorhythmicity, is known as the pacemaker of the heart.

The non-SA nodal autorhythmic tissues are latent pacemakers.

## The spread of cardiac excitation is coordinated to ensure efficient pumping (page 214).

For efficient cardiac function, the spread of excitation should satisfy two criteria:
(1) Atrial excitation and contraction should by complete before the onset of ventricular contraction.  (2) Excitation of cardiac muscle fibers should be coordinated to ensure that each heart chamber contracts as a unit to accomplish efficient pumping.

The normal spread of cardiac excitation is carefully orchestrated to assure that these criteria are met and the heart functions efficiently.

An action potential originating in the SA node first spreads throughout both atria, primarily from cell to cell via gap junctions.

The interatrial pathway extends from the SA node within the right atrium to the left atrium.

This pathway ensures that both atria become depolarized to contract more or less simultaneously.

The internodal pathway extends from the SA node to the AV node.

The internodal conduction pathway directs the spread of an action potential originating at the SA node to the AV node to ensure sequential contraction of the ventricles following atrial contraction.

The action potential is conducted relatively slowly through the AV node.

The slowness is advantageous because it allows time for complete ventricular filling to occur.

The impulse rapidly travels down the bundle of His and throughout the ventricular myocardium via the Purkinje fibers.

The ventricular conduction system is more highly organized and more important than the interatrial and internodal conduction pathways.

The rapid conduction of the action potential down the bundle of His and its swift, diffuse distribution throughout the Purkinje network lead to almost simultaneous activation of the ventricular myocardial cells in both ventricular chambers, which ensures a single, smooth coordinated contraction that can efficiently eject blood into both the systemic and pulmonary circulations at the same time.

## The action potential of contractile cardiac muscle cells shows a characteristic plateau (page 215).

Unlike autorhythmic cells, the membrane of contractile cells remains essentially at rest at about -90mV until excited by electrical activity propagated from the pacemaker.

Once the membrane is excited, an action potential is generated.

During the rising phase of the action potential, the membrane potential rapidly becomes reversed to a positive value of +30mV as a result of an explosive sodium influx.

The membrane potential is maintained at this positive level for several hundred milliseconds, producing a plateau phase of the action potential.

The rapid falling phase of the action potential is accomplished by potassium leaving the cells.

The mechanism by which an action potential in a cardiac muscle fiber brings about contraction of that fiber is quite similar to the excitation - contraction coupling process of skeletal muscle.

In contrast to skeletal muscle cells, calcium also diffuses into the cytosol across the plasma membrane.

Entering calcium triggers even further release of calcium from the sarcoplasmic reticulum.

This extra supply of calcium is not only the major factor responsible for the prolongation of the cardiac action potential but is also responsible for the subsequent lengthening of the period of cardiac contraction.

## Tetanus of cardiac muscle is prevented by a long refractory period (page 215).

During the refractory period, which occurs immediately after the initiation of an action potential, an excitable membrane's responsiveness is totally abolished.

Cardiac muscle has a long refractory period.

Consequently, cardiac muscle cannot be restimulated until contraction is almost over, making summation of contractions and tetanus of cardiac muscle impossible.

## The ECG is a record of the overall spread of electrical activity through the heart (page 216).

An ECG is a recording of that portion of the electrical activity induced in the body fluids by the cardiac impulse that reaches the surface of the body, not a direct recording of the actual electrical activity of the heart.

The ECG is a complex recording representing the overall spread of activity throughout the heart during depolarization and repolarization.

The ECG is not a recording of a single action potential in a single cell.

The recording represents comparisons in voltage detected by electrodes at two different points on the body surface, not the actual potential.

To provide standard comparisons, ECG records routinely consist of twelve conventional electrode systems or leads.

## Various components of the ECG record can be correlated to specific cardiac events (page 218).

A normal ECG exhibits three distinct wave forms: the P wave, the QRS complex, and the T wave.

The P wave represents atrial depolarization.

The QRS complex represents ventricular depolarization.

The T wave represents ventricular repolarization.

The P wave occurs when the impulse spreads across the atria.

The electrical activity associated with atrial repolarization normally occurs simultaneously with ventricular depolarization and is marked by the QRS complex.

The P wave is much smaller than the QRS complex because the atria have a much smaller muscle mass than the ventricles.

There are three times when no current is flowing in the heart musculature and the ECG remains at baseline: (1) During the AV nodal delay, (2) the ST segment, and (3) the TP interval.

## The ECG can be useful in diagnosing abnormal heart rates, arrhythmias, and damage of heart muscle (page 218).

Because electrical activity triggers mechanical activity, abnormal electrical patterns are usually accompanied by abnormal contractile activity of the heart.

The principle deviations from normal that can be ascertained through electrocardiograph are as follows: tachycardia, bradycardia, arrhythmia.

Abnormal ECG waves are also important in the recognition and assessment of cardiac myopathies, such as myocardial ischemia, necrosis and acute myocardial infarction.

# Mechanical Events of the Cardiac Cycle (Text page 220)

The heart alternately contracts to empty and relaxes to fill (page 220).

The cardiac cycle consists of alternate periods of systole and diastole.

The atria and ventricles go through separate cycles of systole and diastole.

Because of the continuous inflow of blood from the venous system into the atrium, atrial pressure exceeds ventricular pressure even though both chambers are relaxed.

The AV valve opens and blood flows directly from the atrium into the ventricle throughout ventricular diastole.

Ventricular volume slowly continues to rise even before atrial contraction takes place.

Late in ventricular diastole, the SA node fires.

The impulse spreads throughout the atria (P wave).

Atrial depolarization brings about atrial contraction.

Throughout atrial contraction, atrial pressure still slightly exceeds ventricular pressure, so the AV valve remains open.

Ventricular diastole ends at the onset of ventricular contraction.

The volume of blood in the ventricle at the end of diastole is known as the end-diastolic volume.

Following atrial excitation, the impulse passes through the AV node and specialized conducting system to excite the ventricle.

Simultaneously, atrial contraction is occurring.

By the time ventricular activation is complete, atrial contraction is already accomplished.

The QRS complex represents this ventricular excitation, which induces ventricular contraction.

As ventricular contraction begins, ventricular pressure immediately exceeds atrial pressure.

The backward pressure differential forces the AV valve closed.

There is a brief period of time between closure of the AV valve and opening of the aortic valve when the ventricle remains a closed chamber.

This interval is termed the period of isovolumetric ventricular contraction.

When ventricular pressure exceeds aortic pressure, the aortic valve is forced open and ejection of blood begins.

Ventricular systole includes both the period of isovolumetric contraction and the ventricular ejection phase.

About half of the blood contained within the ventricle at the end of diastole is pumped out during the subsequent systole.

The amount of blood pumped out of each ventricle with each contraction is known as the stroke volume, the difference between the volume of blood in the ventricle before contraction and the volume after contraction.

T wave signifies ventricular repolarization occurring at the end of ventricular systole.

As the ventricle starts to relax, the aortic valve closes.

Closure of the aortic valve produces a disturbance or notch on the aortic pressure curve known as the dicrotic notch.

All valves are once again closed for a brief period of time known as isovolumetric ventricular relaxation.

When the ventricular pressure falls below the atrial pressure, the AV valve opens and ventricular filling occurs once again.

Two heart sounds associated with valve closures can be heard during the cardiac cycle (page 222).

> Two major heart sounds normally can be heard during the cardiac cycle when listening with a stethoscope.
>
> The first sound is associated with the closure of the AV valve, whereas the second sound is associated with the closure of the semilunar valves.
>
> Closure of the AV valve occurs at the onset of ventricular contraction.
>
> Closure of the semilunar valves occurs at the onset of ventricular relaxation.

Turbulent blood flows produces heart murmurs (page 222).

> Abnormal heart sounds, or murmurs, are usually associated with cardiac disease.
>
> Blood normally flows in a laminar fashion and does not produce any sound.
>
> When blood flow becomes turbulent, however, a sound can be heard and is due to vibrations created in the surrounding structures by the turbulent flow.
>
> The most common cause of turbulence is valve malfunction, either a stenotic (a stiff, narrowed valve) or an insufficient valve (one that cannot close completely).
>
> The valve involved and the type of defect can usually by detected by the location and timing of the murmur.
>
> A murmur occurring between the first and second heart sounds signifies a systolic murmur.
>
> A diastolic murmur occurs between the second and first heart sounds.
>
> The sound of the murmur characterizes it as either a stenotic (whistling) murmur or an insufficient (swishy) murmur.

## Cardiac Output and Its Control (Text page 223)

Cardiac output depends on the heart rate and stroke volume (page 223).

> Cardiac output is the volume of blood pumped by each ventricle per minute.
>
> The two determinants of cardiac output are heart rate and stroke volume.

Heart rate is determined primarily by autonomic influences on the SA node (page 223).

> The heart is innervated by both divisions of the autonomic nervous system, which can modify the rate of contraction, even though nervous stimulation is not required to initiate contraction.
>
> The parasympathetic nerve to the heart; the vagus nerve, primarily supplies the atrium, especially the SA and AV nodes.
>
> There is no significant parasympathetic innervation to the ventricles.
>
> The cardiac sympathetic nerves also supply the atria, including the SA and AV nodes, and richly innervate the ventricles as well.
>
> The parasympathetic nervous system's influence on the SA node is to decrease the heart rate.
>
> Parasympathetic influence on the AV node decreases the node's excitability, prolonging the AV nodal delay.
>
> Parasympathetic stimulation of the atrial contractile cells shortens the action potential, that is, the atrial contraction is weakened.
>
> The sympathetic nervous system speeds up the heart rate through its effect on the pacemaker tissue.
>
> Sympathetic stimulation of the AV node reduces the AV nodal delay.

Similarly, sympathetic stimulation speeds up the spread of the action potential throughout the specialized conducting pathway.

Sympathetic stimulation increases contractile strength so that the heart beats more forcefully and squeezes out more blood.

## Stroke volume is determined by the extent of venous return and by sympathetic activity (page 225).

Two types of controls influence stroke volume: (1) intrinsic control related to the extent of venous return and (2) extrinsic control related to the extent of sympathetic stimulation of the heart.

## Increased end-diastolic volume results in increased stroke volume (page 225).

The direct correlation between end-diastolic volume and stroke volume constitutes the intrinsic control of stroke volume, which refers to the heart's inherent ability to vary the stroke volume.

For cardiac muscle, the resting cardiac muscle fiber length is less than optimal length.

An increase in cardiac muscle fiber length, by moving closer to the optimal length, increases the contractile tension of the heart on the following systole.

The main determinant of cardiac muscle fiber length is the degree of diastolic filling.

This intrinsic relationship between end-diastolic volume and stroke volume is known as the Frank-Starling law of the heart.

Stated simply, the law says that the heart normally pumps all the blood returned to it.

## The contractility of the heart is increased by sympathetic stimulation (page 226).

Stroke volume is also subject to extrinsic control.

Sympathetic stimulation and epinephrine enhance the heart's contractility.

Sympathetic stimulation increase stroke volume by enhancing venous return.

Sympathetic stimulation constricts the veins, which squeezes more blood forward from the veins to the heart, increasing the end-diastolic volume.

## The contractility of the heart is decreased in heart failure (page 227).

Heart failure refers to the inability of the cardiac output to keep pace with the body's demands for supplies and removal of wastes.

Heart failure may occur for a variety of reasons, but the two most common are (1) damage to the heart muscle and (2) prolonged pumping against a chronically elevated blood pressure.

The prime defect in heart failure is a decrease in cardiac contractility.

A failing heart will pump out a smaller stroke volume than a normal healthy heart.

Two major compensatory measures help restore the stroke volume to normal; sympathetic activity to the heart is reflexly increased, and when cardiac output is reduced, the kidneys, in a compensatory attempt to improve their reduced blood flow, retain extra salt and water in the body during urine formation to expand the blood volume.

## Nourishing the Heart Muscle (Text page 229)

The heart receives most of its own blood supply through the coronary circulation during diastole (page 229).

> The heart muscle must receive blood through blood vessels, specifically by means of the coronary circulation.
>
> The coronary arteries branch from the aorta just beyond the aortic valve, and the coronary veins empty into the right atrium.
>
> Most coronary arterial flow occurs during diastole.
>
> Coronary blood flow is adjusted primarily in response to changes in the heart's oxygen requirements.

Atherosclerotic coronary artery disease can deprive the heart of essential oxygen (page 229).

> Coronary artery disease can cause myocardial ischemia by the three following mechanisms: (1) profound vascular spasm of the coronary arteries; (2) the formation of atherosclerotic plaques; and (3) thromboembolism.
>
> Vascular spasm is an abnormal spastic constriction that transiently narrows the coronary vessels.
>
> Atherosclerosis is a progressive, degenerative arterial disease that leads to occlusion of affected vessels.
>
> Potential complications of coronary atherosclerosis are angina pectoris and thromboembolism.

The amount of "good" cholesterol versus "bad" cholesterol in the blood is linked to atherosclerosis (page 230).

> There are three major lipoproteins, named for their density of protein as compared to lipid: (1) high-density lipoproteins (HDL); (2) low-density lipoproteins (LDL); and (3) very-low-density lipoproteins (VLDL).
>
> Cholesterol carried in LDL complexes has been termed "bad" cholesterol, because cholesterol is transported to the cells, including those lining the blood-vessel walls, by means of LDL.
>
> In contrast, cholesterol carried in HDL complexes has been dubbed "good" cholesterol, because HDL removes cholesterol from the cells and transports it to the liver for partial elimination from the body.
>
> The liver has a primary role in determining total blood cholesterol levels, and the interplay between LDL and HDL determines the traffic flow of cholesterol between the liver and the individual cells of the body.
>
> Evidence suggests that the propensity toward developing atherosclerosis substantially increases with elevated levels of LDL.
>
> Elevated levels of HDL are associated with a low incidence of atherosclerotic heart disease.
>
> The ingestion of polyunsaturated fatty acids, the predominant fatty acids of most plants, tends to reduce blood cholesterol levels by enhancing the elimination of both cholesterol and cholesterol-derived bile salts in the feces.

## KEY TERMS

Angina pectoris-p229
Arrhythmia-p218
Atherosclerosis-p229
Atheromas-p229
Autorhythmicity-p211
Bradycardia-p218
Bundle of His-p212
Cardiac myopathies-p218
Cardiac output-p223

Diastole-p220
Electrocardiogram-p216
Endocardium-p209
Epicardium-p209
Fibrillation-p214
Myocardial infarction-p218
Myocardial ischemia-p218
Myocardium-p209
Necrosis-p218

Percarditis-p211
Purkinje fibers-p212
P wave-p218
QRS complex-p218
Stenotic valve-p222
Stroke volume-p220
Systole-p220
Tachycardia-p218
T wave-p218

## REVIEW EXERCISES

**True - False** (Answers on page A-16)

_____ 1. Ninety-nine percent of the cardiac-muscle cells are autorhythmic cells.

_____ 2. The AV node, which normally exhibits the fastest rate of autorhythmicity is known as the pacemaker of the heart.

_____ 3. An action potential originating in the SA node first spreads throughout both atria primarily from cell to cell via gap junctions.

_____ 4. The internodal conduction pathway directs the spread of an action potential originating at the SA node to the AV node to ensure sequential contraction of the ventricles following atrial contraction.

_____ 5. The ventricular conduction system is more highly organized and more important than the interatrial and internodal conduction pathways.

_____ 6. Cardiac muscle has a short refractory period.

_____ 7. A normal ECG exhibits three distinct wave forms:  the P wave, the QRS complex, and the T wave.

_____ 8. The QRS complex represents atrial depolarization.

_____ 9. The two determinants of cardiac output are heart rate and stroke volume.

_____ 10. The parasympathetic nervous system's influence on the SA node is to increase the heart rate.

_____ 11. The parasympathetic nervous system stimulation of the AV node reduces the AV nodal delay.

_____ 12. Sympathetic stimulation increases stroke volume by enhancing venous return.

_____ 13. The prime defect in heart failure is a decrease in cardiac contractility.

_____ 14. Cholesterol carried in HDL complexes has been termed "bad" cholesterol.

_____ 15. The liver has a primary role in determining total blood cholesterol levels.

_____ 16. Elevated levels of HDL are associated with a low incidence of atherosclerotic heart disease.

_____ 17. The cardiac cycle consists of alternate periods of systole and diastole.

_____ 18. Early in ventricular diastole, the SA node reaches threshold and fires.

_____ 19. Atrial depolarization brings about atrial contraction.

_____ 20. Ventricular systole includes both the period of isovolumetric contraction and the ventricular ejection phase.

_____ 21. The first heart sound is associated with closure of the AV valve.

_____ 22. Laminar flow produces a faint sound.

_____ 23. The sound of the murmur characterizes it as either a stenotic murmur or an insufficient murmur.

_____ 24. There are no gap junctions between the atrial and ventricular contractile cells.

_____ 25. Either all the cardiac-muscle fibers contract or none of them do.

_____ 26. Blood serves as the transport medium.

_____ 27. Pulmonary circulation consist of a circuit of vessels carrying blood between the heart and organ systems.

_____ 28. Anatomically the heart is a single organ, though it has a right and left side, they still function as a single pump.

_____ 29. The left side of the heart pumps blood into the pulmonary circulation.

_____ 30. Both sides of the heart simultaneously pump equal amounts of blood.

_____ 31. The systemic circulation is a low-pressure, low-resistance system.

_____ 32. Blood flows through the heart in one fixed direction from veins to atria to ventricles to arteries.

_____ 33. Cardiac muscle is capable of generating action potentials without any nervous stimulation.

## Fill in the Blank (Answers on page A-16)

34. _____ is an abnormal spastic constriction that transiently narrows the coronary vessels.

35. _____ is a progressive, degenerative arterial disease that leads to occlusion of affected vessels.

36. An abnormal clot attached to a vessel wall is known as a _____. A freely floating clot is known as _____.

37. _____ exists when small terminal branches from adjacent blood vessels nourish the same area.

38. There are three major lipoproteins, (1) _____ which contain the most protein and least cholesterol, (2) _____ which contain less protein and more cholesterol, and (3) _____ which contain the least protein and most lipid, but the lipid carried is neutral fat, not cholesterol.

39. Random, uncoordinated excitation and contraction of the cardiac cells is known as _____.

40. The _____ pathway extends from the SA node within the right atrium to the left atrium.

41. The _____ represents ventricular repolarization.

42. A rapid heart rate of more than 100 beats per minute is known as _____.
43. A slow heart rate of fewer than 60 beats per minute is known as _____.
44. The heart has a broad base at the top and tapers to a pointed tip known as the _____ at the bottom.
45. CPR is short for _____.
46. _____ pump blood from the heart.
47. The vessels that return blood from the tissues to the atria are _____.
48. The large artery carrying blood away from the left ventricle is the _____.
49. The right AV valve is also called the _____.
50. The _____ is a thin external membrane covering the heart.
51. The non-SA nodal autorhythmic tissues are _____.
52. _____ is the volume of blood pumped by each ventricle per minute.
53. The difference between the cardiac output at rest and the maximum volume of blood the heart is capable of pumping per minute is known as the _____.
54. The direct correlation between end-diastolic volume and stroke volume constitutes the _____ of stroke volume, which refers to the heart's inherent ability to vary the stroke volume.
55. The intrinsic relationship between end-diastolic volume and stroke volume is known as the _____.
56. _____ refers to the inability of the cardiac output to keep pace with the body's demands for supplies and removal of wastes.
57. The volume of blood in the ventricle at the end of diastole is known as the _____ _____.
58. When all the valves are closed and no blood can enter or leave the ventricle during this time, this interval is termed the period of _____.
59. The amount of blood remaining in the ventricle at the end of systole when ejection is complete is known as the _____.
60. Abnormal heart sounds are called _____.
61. An abnormal heart sound occurring between the first and second heart sounds signifies a _____.

**Matching:** Match the division of the autonomic nervous system to the effect it exerts on the heart or structures that influence the heart (Answers on page A-17)

_____ 62. Has no effect on the adrenal medulla

_____ 63. Increases contractility of the atrial muscle and strengths contraction

_____ 64. Decreases excitability; increases AV nodal delay

_____ 65. Has no effect on the ventricular conduction pathway

_____ 66. Increases rate of depolarization to threshold of SA node; increases heart rate

_____ 67. Promotes adrenomedullary secretion of epinephrine

_____ 68. Increases contractility of the ventricular muscle; strengthens contraction

_____ 69. Increases venous return

_____ 70. Decreases contractility of atrial muscle

_____ 71. Increases excitability; decreases AV nodal delay

a. sympathetic

b. parasympathetic

**Multiple Choice** (Answers on page A-17)

_____ 72. This cardiac disorder is most often characterized by a whistling murmur.
   a. myocardial infarction
   b. myocardial ischemia
   c. atherosclerosis
   d. congestive heart failure
   e. valvular stenosis

_____ 73. This cardiac complication can be caused by prolonged pumping against a chronically increased afterload.
   a. myocardial infarction
   b. myocardial ischemia
   c. atherosclerosis
   d. congestive heart failure
   e. valvular stenosis

_____ 74. This chest pain is associated with myocardial ischemia.
   a. myocardial nociceptosis
   b. angina pectoris
   c. atheroma
   d. valvular stenosis
   e. heart burn

_____ 75. This is a noncancerous tumor of smooth-muscle cells within the blood-vessel walls.
   a. myocardial infarction
   b. myocardial ischemia
   c. atheroma
   d. fibroblastoma
   e. glioma

_____ 76. This cardiac disorder can be caused by profound vascular spasm of the coronary arteries.
   a. myocardial infarction
   b. myocardial ischemia
   c. atherosclerosis
   d. congestive heart failure
   e. valvular stenosis

_____ 77. This is a progressive, degenerative arterial disease that leads to occlusion of affected vessels.
   a. myocardial infarction
   b. myocardial ischemia
   c. atherosclerosis
   d. congestive heart failure
   e. valvular stenosis

_____ 78. This heart disorder can be caused by thromboembolism.
   a. myocardial infarction
   b. myocardial ischemia
   c. atherosclerosis
   d. congestive heart failure
   e. valvular stenosis

_____ 79. This disease has a close relationship with cholesterol.
   a. myocardial infarction
   b. myocardial ischemia
   c. atherosclerosis
   d. congestive heart failure
   e. valvular stenosis

_____ 80. This disease is characterized by abnormally large amounts of blood being dammed up in the venous system.
   a. myocardial infarction
   b. myocardial ischemia
   c. atherosclerosis
   d. congestive heart failure
   e. valvular stenosis

_____ 81. This condition is commonly known as a heart attack.
a. myocardial infarction
b. myocardial ischemia
c. atherosclerosis
d. congestive heart failure
e. valvular stenosis

## Points to Ponder
1. Congenital heart defects are routinely corrected through the miracles of modern surgery. What happens to the number of people having such disorders as a result of these surgical techniques? What is the solution to this problem?
2. After carefully studying the sections of this chapter involving cholesterol and lipids, why do you suppose a good physician checks the triglycerides in the blood?

## Clinical Perspectives
1. With respect to the heart why do people generally sleep in a prone position?
2. What happens to the heart in these two situations?
   A. You sense you are about to sneeze so you close your mouth and hold your nose to muffle the sneeze. Then you sneeze.
   B. You are extremely constipated. In an effort to have a bowel movement you strain very hard.
3. Why do you suppose the sympathetic nervous system has an effect on the adrenal medulla while the parasympathetic has no effect at all?
4. Why is the nitroglycerine tablet placed under the tongue in cardiac patients?

# BLOOD VESSELS AND

# BLOOD PRESSURE

**8**

## CHAPTER OVERVIEW

The circulatory system aids in the performance of many vital functions. The circulatory contribution to all these functions is flow. Flow depends on pressure and resistance.

The circulatory system includes arteries, arterioles, capillaries, venules, and veins. It is the flow of blood through these vessels that is so important to homeostasis. For this system to function pressure and resistance are essential. The heart and the blood vessels provide the pressure. The pressure is monitored by baroreceptors which provide the CNS with the information necessary to control the blood flow.

Arteries and veins play a major role in the gross transport of blood. The cardiac output is distributed throughout the body in a highly variable pattern. In this capacity it is the arterioles and capillaries that play the major role. Through changes in resistance the arterioles provide the mechanism for the distributional changes. The distribution of cardiac output is adjusted to maintain homeostasis.

Consider all the substances released into or removed from the blood. Many of these molecules have typical cells as their destinations, but most are being transported to specific tissues. The capillaries are the sites of exchange. Arterioles, which respond to intrinsic and extrinsic controls, provide the proper pressure for the exchanges to occur. The circulatory system, like the other body systems, functions to maintain homeostasis. Again, all levels of organization are performing to maintain homeostasis for the survival of cells.

# CHAPTER OUTLINE

## Introduction (Text page 238)

Materials are transported within the body by the blood as it is pumped through the blood vessels (page 238).

The majority of the blood cells are not in direct contact with the external environment, yet these cells must make exchanges with the environment, such as picking up oxygen and nutrients and eliminating wastes.

All blood pumped by the right side of the heart passes through the lungs for oxygen pickup and carbon dioxide removal.

The blood pumped by the left side of the heart is parceled out in various proportions to the systemic organs through a parallel arrangement of vessels that branch from the aorta.

Blood is constantly "reconditioned" so that its composition remains relatively constant despite an ongoing drain of supplies to support metabolic activities and the continual addition of wastes from the tissues.

The organs that recondition the blood normally receive substantially more blood than is necessary to meet their basic metabolic needs so that they can perform homeostatic adjustments on the blood.

Reconditioning organs can withstand temporary reductions in blood flow.

In contrast, the brain can least tolerate a disruption in its blood supply.

Blood flow through vessels depends on the pressure gradient and vascular resistance (page 239).

Arteries carry blood from the heart.

When an artery reaches the organ it is supplying, it branches into numerous arterioles.

Arterioles branch further within the organs into capillaries.

Capillaries rejoin to form small venules, which further merge to form small veins.

The small veins unite to form larger veins.

The flow rate of blood through a vessel is directly proportional to the pressure gradient and inversely proportional to vascular resistance.

The pressure gradient--the difference in pressure between the beginning and end of a vessel--is the main driving force for flow through the vessel.

The greater the pressure gradient forcing blood through a vessel, the greater the rate of flow through that vessel.

Resistance is a measure of the hindrance to blood flow through a vessel caused by friction between the moving fluid and the stationary vascular walls.

Resistance to blood flow depends on three factors: (1) viscosity of the blood; (2) vessel length, and (3) vessel radius.

## Arteries (Text page 242)

Arteries serve as rapid-transit passageways to the tissues and as a pressure reservoir (page 242).

Arteries are specialized to serve as rapid-transit passageways for blood from the heart to the tissues and to act as a pressure reservoir to provide the driving force for blood when the heart is relaxing.

Capillary flow does not fluctuate between cardiac systole and diastole; blood flow is continuous through the capillaries supplying the tissues.

When the heart relaxes and ceases pumping blood into the arteries, the stretched arterial walls passively recoil and push the excess blood contained in the arteries into the  vessels downstream.

### Arterial pressure fluctuates in relation to ventricular systole and diastole (page 244).

The maximum pressure exerted in the arteries when the blood is ejected into them during systole, the systolic pressure, averages 120 mm Hg.

The minimum pressure within the arteries when blood is draining off into the remainder of the vessels during diastole,  the diastolic pressure, averages 80 mm Hg.

### Blood pressure can be indirectly measured by using a sphygmomanometer (page 244).

It is convenient and reasonably accurate to measure the pressure indirectly through the use of a sphygmomanometer, an externally applied inflatable cuff attached to a pressure gauge.

During the determination of blood pressure, a stethoscope is placed over the brachial artery.

At the onset of a blood-pressure determination, the cuff is inflated to a pressure greater than systolic blood pressure so that the brachial artery collapses.

The pressure in the cuff is gradually reduced.

The highest cuff pressure at which the first sound can be heard is indicative of the systolic blood pressure.

The highest cuff pressure at which the last sound can be  detected is indicative of the diastolic pressure.

The pressure difference between systolic and diastolic pressures is known as the pulse pressure.

### Mean arterial pressure is the main  driving force for blood flow (page 244).

The mean arterial pressure is the average pressure responsible for driving blood forward into the tissues throughout the cardiac cycle.

The mean arterial pressure averages 93 mm Hg.

Arterial pressure is essentially the same throughout the arterial tree.

## Arterioles (Text page 246)

### Arterioles are the major resistance vessels (page 246).

The arterioles are the major resistance vessels in the vascular tree, even though the capillaries have smaller radii than the arterioles.

The radii of arterioles supplying individual organs can be adjusted independently to determine the distribution of cardiac output and to regulate arterial blood pressure.

Arteriolar walls have a thick layer of smooth muscle that is richly innervated by sympathetic nerve fibers.  The smooth-muscle layer runs circularly around the arteriole so that when it contracts, the vessel's circumference becomes smaller, thus increasing resistance and decreasing the flow through that vessel.

Local (intrinsic) controls and extrinsic controls influence the level of contractile activity in arteriolar smooth muscle.

Local control of arteriolar radius is important in determining distribution of cardiac output so that blood flow is matched with the tissues' metabolic needs (page 246).

Blood is delivered to all tissues at the same mean arterial pressure.

The distribution of cardiac output can be varied by differentially adjusting arteriolar resistance in the various vascular beds.

Local controls are changes within a tissue that alter the radii of the vessels and hence adjust blood flow through the tissue by directly affecting the smooth muscle of the tissue's arterioles.

Local influences may be either chemical or physical in nature.

Local chemical changes produce dilation without involving nerves or hormones.

The following local chemical factors produce relaxation of arteriolar smooth muscles: decreased oxygen, increased carbon dioxide, increased acid, increased potassium, increased osmolarity, adenosine release, and prostaglandin release.

The single layer of specialized epithelial cells that line the lumen of all blood vessels, the endothelial cells, release nitric oxide.

Nitric oxide causes relaxation of arteriolar smooth muscle.

By dilating the arterioles of the penis, nitric oxide is the direct mediator of penile erection.

Macrophages produce nitric oxide, which they use as "chemical warfare" against bacteria and cancer cells.

Nitric oxide interferes with platelet function and blood clotting at sites of vessel damage.

Nitric oxide serves as a neurotransmitter in the brain and elsewhere.

Nitric oxide plays a role in the changes underlying memory.

The endothelial cells release other important chemicals, such as endothelin, which bring about vasoconstriction by causing arteriolar smooth-muscle contraction.

Histamine is another local chemical mediator that influences arteriolar smooth muscle, but it is not released in response to local metabolic changes and is not derived from the endothelial cells.

Histamine is synthesized and stored within special connective-tissue cells in many tissues and in certain types of circulating white blood cells.

When tissues are injured or during allergic reactions, histamine is released in the damaged region.

Histamine is the major cause of vasodilation in an injured area.

Local physical influences on arterioles include heat or cold.

Extrinsic control of arteriolar radius is primarily important in the regulation of arterial blood pressure (page 250).

Extrinsic control of arteriolar radius includes both neural and hormonal influences, with the effects of the sympathetic nervous system being the most important.

Sympathetic nerve fibers supply arteriolar smooth muscle everywhere except in the brain.

A certain level of ongoing sympathetic activity contributes to vascular tone.

Increased sympathetic activity produces generalized arteriolar vasoconstriction, whereas decreased sympathetic activity leads to generalized arteriolar vasodilation.

The extent to which each organ actually receives blood flow is determined by local arteriolar adjustments that override the sympathetic constrictor effect.

The main region of the brain responsible for adjusting sympathetic output to the arterioles is the cardiovascular control center in the medulla of the brain stem.

Several hormones also extrinsically influence arteriolar radius including epinephrine, vasopressin and angiotension II.

# Capillaries (Text page 251)

## Capillaries are ideally suited to serve as sites of exchange (page 251).

Capillaries, the sites of exchange of materials between the blood and tissues, branch extensively to bring blood within the reach of every cell.

Exchange of materials across capillary walls is accomplished by the process of diffusion.

Diffusing molecules have only a short distance to travel between the blood and surrounding cells because of the thin capillary wall and small capillary diameter, coupled with the close proximity of each and every cell to a capillary.

Because capillaries are distributed in such incredible numbers, a tremendous total surface area is available for exchange.

Diffusion is enhanced because blood flows more slowly in the capillaries than elsewhere in the circulatory system.

## Water-filled pores in the capillary wall permit passage of small, water-soluble substances that cannot cross the endothelial cells themselves (page 254).

Diffusion across capillary walls also depends on the walls permeability to the materials being exchanged.

In most capillaries, narrow, water-filled clefts, or pores, are present at the junctions between the cells.

The size of the capillary pore varies from organ to organ.

In response to appropriate signals, the endothelial cells can adjust themselves to vary the size of the pores.

Vesicular transport also plays a limited role in the passage of materials across the capillary wall.

## Diffusion across the capillary wall is important in solute exchange (page 254).

Exchanges are not made directly between blood and the tissue cells.

Cells exchange materials directly with the interstitial fluid, the type and extent of exchange being governed by the properties of the cellular plasma membranes.

Passive diffusion down concentration gradients is the primary mechanism for exchange of individual solutes.

## Bulk flow across the capillary wall is important in extracellular fluid distribution (page 256).

Bulk flow is the process whereby a volume of protein-free plasma filters out of the capillary, mixes with the surrounding interstitial fluid, and is subsequently reabsorbed.

Bulk flow occurs because of differences in the hydrostatic and colloid osmotic pressures between the plasma and interstitial fluid.

Bulk flow does not play an important role in the exchange of individual solutes between blood and tissues, because the quantity of solutes moved across the capillary wall by bulk flow is extremely small compared to the much larger transfer of solutes by diffusion.

Bulk flow plays an extremely important role in regulating the distribution of ECF between plasma and interstitial fluid.

The lymphatic system is an accessory route by which interstitial fluid can be returned to the blood (page 258).

Even under normal circumstances, slightly more fluid is filtered out of the capillaries into the interstitial fluid than is reabsorbed from the interstitial fluid back into the plasma.

The extra fluid filtered out as a result of this filtration-reabsorption imbalance is picked up by the lymphatic system.

Small, blind-ended terminal lymph vessels (lymphatic capillaries) permeate almost every tissue of the body.

Once interstitial fluid enters a lymphatic vessel, it is called lymph.

Lymphatic capillaries converge to form larger and larger lymph vessels, which eventually empty into the venous system.

The most important functions of the lymphatic system are as follows: (1) the return of excess filtered fluid to the blood, (2) the defense against disease provided by phagocytic cells in the lymph nodes, (3) the transport of absorbed fat; and (4) the return of filtered protein to the blood.

Edema occurs when too much interstitial fluid accumulates (page 260).

Swelling of the tissues because of excess interstitial fluid is known as edema.

The cause of edema can be grouped into four general categories: (1) a reduced concentration of plasma proteins; (2) an increased permeability of the capillary walls; (3) an increased venous pressure; and (4) the blockage of lymph vessels.

Whatever the cause of edema, an important consequence is a reduction in exchange of materials between the blood and cells.

# Veins (Text page 261)

Veins serve as a blood reservoir as well as passageways back to the heart (page 261).

Because the total cross-sectional area of the venous system gradually decreases as smaller veins converge into progressively fewer but larger vessels, the velocity of blood flow increases as the blood approaches the heart.

Systemic veins serve as a blood reservoir.

When the stored blood is needed, such as during exercise, extrinsic factors drive the extra blood from the veins to the heart.

A delicate balance exists between the capacity of the veins, the extent of venous return, and the cardiac output.

Venous return is enhanced by a number of extrinsic factors (page 262).

Changes in venous capacity directly influence the magnitude of venous return, which in turn is an important determinant of effective circulating blood volume.

Sympathetic stimulation produces venous vasoconstriction, which modestly elevates venous pressure; this in turn, increases the pressure gradient to drive more blood from the veins into the right atrium.

Venous vasoconstriction enhances venous return by decreasing venous capacity.

Many of the large veins in the extremities lie between skeletal muscles so that when the muscles contract, the veins are compressed.

External venous compression decreases venous capacity and increases venous pressure, in effect squeezing fluid contained in the veins forward toward the heart.

This pumping action, known as the skeletal muscle pump, is one way by which extra blood stored in the veins is returned to the heart during exercise.

Increased muscular activity pushes more blood out of the veins and into the heart.

Blood can only be driven forward because the large veins are equipped with one-way valves.

These valves permit blood to move forward toward the heart but prevent it from moving back toward the tissues.

Varicose veins occur when the venous valves become incompetent and can no longer support the column of blood above them.

The most serious consequence of varicosed veins is the possibility of abnormal clot formation in the sluggish, pooled blood.

As a result of respiratory activity, the pressure within the chest cavity averages 5 mm Hg less than atmospheric pressure.

The pressure difference between the lower veins and the chest veins squeezes blood from the lower veins to the chest veins, promoting venous return.

During ventricular contraction, the AV valves are drawn downward, enlarging the atrial cavities, which drops the atrial pressure to below 0 mm Hg, thus enhancing venous return.

# Blood Pressure (Text page 265)

Regulation of mean arterial blood pressure is accomplished by controlling cardiac output, total peripheral resistance and blood volume (page 265).

Mean arterial blood pressure is the main driving force propelling blood to the tissues. The mean arterial blood pressure must be high enough to ensure sufficient driving pressure, but not be so high that it creates extra work for the heart and increases the risk of vascular damage and possible rupture of small blood vessels.

Mean arterial pressure is constantly monitored by baroreceptors within the circulatory system.

Short-term adjustments are accomplished by alterations in cardiac output and total peripheral resistance, mediated by means of autonomic nervous -system influences on the heart, veins, and arterioles.

Long-term control involves adjusting total blood volume by restoring normal salt and water balance through mechanisms that regulate urine output and thirst.

The baroreceptor reflex is the most important mechanism for short-term regulation of blood pressure (page 267).

Any change in mean blood pressure triggers an autonomically mediated baroreceptor reflex that influences the heart and blood vessels to adjust cardiac output and total peripheral resistance in an attempt to restore blood pressure to normal.

The carotid-sinus and aortic-arch baroreceptors are sensitive to changes in both mean arterial pressure and pulse pressure.

The integrating center that receives the afferent impulses about the status of arterial pressure is the cardiovascular control center.

The cardiovascular control center alters the ratio between sympathetic and para-sympathetic activity to the effector organs.

Hypertension is a serious national public health problem, but its causes are largely unknown (page 268).

> Secondary hypertension occurs secondary to another primary problem such as arteriosclerosis.
>
> The underlying cause is unknown in primary hypertension.
>
> There is a strong genetic tendency to develop primary hypertension, which can be hastened or worsened by contributing factors such as obesity, stress, smoking, and excessive ingestion of salt.
>
> Whatever the underlying defect, once initiated, hypertension appears to be self perpetuating.
>
> Constant exposure to elevated blood pressure predisposes vessel walls to the development of atherosclerosis, which further elevates blood pressure.
>
> The baroreceptors do not respond to bring the blood pressure back to normal during hypertension because they adapt or are "reset" to operate at a higher level.
>
> Complications of hypertension include congestive heart failure, strokes, or heart attacks.

Inadequate sympathetic activity is responsible for dizziness or fainting accompanying transient orthostatic hypotension (page 270).

> Hypotension occurs either when there is a disproportion between vascular capacity and blood volume or when the heart is too weak to impart sufficient driving pressure to the blood.
>
> Orthostatic hypotension is a transient hypotensive condition resulting from insufficient compensatory responses to the gravitational shifts in blood that occur when a person moves from horizontal to a vertical position.
>
> Hypotension results from blood pooling in the leg veins.

Circulatory shock can become irreversible (page 270).

> When blood pressure falls so low that adequate blood flow to the tissues can no longer be maintained, the condition known as circulatory shock occurs.
>
> Compensatory mechanisms are often insufficient in the face of substantial fluid loss. Fluid volume must be replaced from the outside through drinking, transfusion, or a combination of both.
>
> A point may be reached at which blood pressure continues to drop rapidly because of tissue damage, despite vigorous therapy.
>
> This condition is termed irreversible shock, in contrast to reversible shock, which can be corrected by compensatory mechanisms and effective therapy.

## KEY TERMS

| | | |
|---|---|---|
| Arteries-p239 | Diastolic pressure-p244 | Resistance-p241 |
| Arterioles-p240 | Endothelin-p250 | Respiratory pump-p264 |
| Baroreceptors-p266 | Flow rate-p240 | Systolic pressure-p244 |
| Bulk flow-p236 | Hypertension-p268 | Vasodilation-p246 |
| Capillaries-p240 | Hypotension-p268 | Veins-p240 |
| Carotid sinus-p267 | Pressure gradient-p240 | Venules-p240 |
| Circulatory shock-p270 | Pulse pressure-p244 | Viscosity-p241 |

# REVIEW EXERCISES

**True - False** (Answers on page A-18)

_____   1. Blood flows more slowly in capillaries than elsewhere in the circulatory system.

_____   2. Diffusion across capillary walls also depends on the wall's permeability to the materials being exchanged.

_____   3. The size of the capillary pores are consistent throughout the body.

_____   4. Solute exchanges are made directly between blood and the tissue cells.

_____   5. Bulk flow occurs because of differences in the hydrostatic and colloid osmotic pressures between the plasma and interstitial fluid.

_____   6. The inability to return filtered protein is a cause of edema.

_____   7. All blood pumped by the left side of the heart passes through the lungs for oxygen pickup and carbon dioxide removal.

_____   8. Reconditioning organs can withstand temporary reductions in blood flow.

_____   9. The greater the pressure gradient forcing blood through a vessel, the lesser the rate of flow through that vessel.

_____   10. Capillary flow fluctuates between cardiac systole and diastole.

_____   11. Systolic pressure averages 120 mm Hg and diastolic pressure averages 80 mm Hg.

_____   12. To determine blood pressure, a stethoscope is placed over the brachial artery.

_____   13. Arterial pressure is essentially the same throughout the arterial tree.

_____   14. Orthostatic hypotension because of emotional stress can also cause dizziness or fainting.

_____   15. Mean arterial blood pressure is the main driving force for propelling blood to the tissues.

_____   16. Long-term adjustments are accomplished by alterations in cardiac output and total peripheral resistance, mediated by means of autonomic nervous-system influences on the heart, veins, and arterioles.

_____   17. The cardiovascular control center alters the ratio between sympathetic and parasympathetic activity to the effector organs.

_____   18. The blood flow to the brain is increased during exercise.

_____   19. Baroreceptors respond and help bring the blood pressure back to normal during hypertension.

_____   20. Arterioles are the major resistance vessels in the vascular tree.

_____   21. Local (intrinsic) controls are changes within a tissue that alter the radii of the vessels and hence adjust blood flow through the tissue by directly affecting the smooth muscle of the tissue's arterioles.

_____   22. Local influences can only be chemical not physical in nature.

_____   23. Nitric oxide causes relaxation of arteriolar smooth muscle.

_____ 24. Histamine is a local chemical mediator that influences arteriolar smooth muscle.

_____ 25. Sympathetic nerve fibers supply arteriolar smooth muscle everywhere including the brain.

_____ 26. Extrinsic control of arteriolar radius includes both neural and hormonal influences.

_____ 27. Systemic veins also serve as a blood reservoir.

_____ 28. Venous vasoconstriction enhances venous return by increasing venous capacity.

_____ 29. Increased muscular activity pushes more blood out of the veins and into the heart.

_____ 30. Blood can be driven forward or backward through the veins.

_____ 31. Abnormal clot formation in the sluggish, pooled blood is a serious risk to a person with varicose veins.

_____ 32. The pressure within the chest cavity averages 11 mm Hg less than atmospheric pressure.

**Fill in the Blank** (Answers on page A-18)

33. _____ is when the blood pressure is above 140/90 mm Hg.

34. _____ in its extreme form is circulatory shock.

35. _____ is a transient hypotensive condition resulting from insufficient compensatory responses to the gravitational shifts in blood that occur when a person moves from a horizontal to a vertical position.

36. _____ shock is when a point is reached where the blood pressure continues to drop rapidly because of tissue damage, despite vigorous therapy.

37. Veins are often referred to as _____.

38. _____ refers to the volume of blood that the veins can accommodate.

39. _____ refers to the volume of blood entering each atrium per minute from the veins.

40. _____ occur when the venous valves become incompetent and can no longer support the column of blood above them.

41. Mean arterial pressure is constantly monitored by _____ within the circulatory system.

42. _____ and _____ are mechanoreceptors sensitive to changes in both mean arterial pressure and pulse pressure.

43. _____ carry blood from the heart.

44. _____ of blood through a vessel is directly proportional to the pressure gradient and inversely proportional to vascular resistance.

45. The maximum pressure exerted in the arteries when blood is ejected into them during systole averages _____.

46. The pressure difference between systolic and diastolic is known as the _____ _____.

47. _____ is the average pressure responsible for driving blood forward into the tissues throughout the cardiac cycle.

48. When a small artery reaches the organ it is supplying, it branches into numerous _____.

49. Exchange of materials across capillary walls is accomplished primarily by the process of _____.

50. _____ is a volume of protein-free plasma that actually filters out of the capillary, mixes with surrounding interstitial fluid, and is subsequently reabsorbed.

51. _____ is the fluid or hydrostatic pressure exerted on the inside of the capillary walls by the blood.

52. The extra fluid filtered out as a result of this ultrafiltration reaborsption imbalance is picked up by the _____.

53. Swelling of the tissues because of excess interstitial fluid is known as _____.

54. _____ is the term applied to such narrowing of a vessel.

55. Arteriolar smooth muscle normally displays a state of partial constriction known as _____.

56. _____ causes local arteriolar vasodilation by inducing relaxation of arteriolar smooth muscle in the vicinity.

57. _____ is a single layer of specialized epithelial cells that line the lumen of all blood vessels.

58. The main region of the brain responsible for adjusting sympathetic output to the arterioles is the _____.

**Matching:** Match the cardiovascular variable to the change as a result of exercise
(Answers on page A-19)

_____ 59. Blood flow to skin

_____ 60. Cardiac output                          a. increases

_____ 61. Venous return                            b. decreases

_____ 62. Blood flow to digestive track           c. unchanged

_____ 63. Blood flow to brain

_____ 64. Stroke volume

_____ 65. Blood flow to bone

_____ 66. Blood flow to heart

_____ 67. Blood flow to kidney

_____ 68. Heart rate

_____ 69. Median arterial blood pressure

_____ 70. Blood flow to active skeletal muscles

_____ 71. Total peripheral resistance

_____ 72. Stored blood

**Multiple Choice** (Answers on page A-19)

_____ 73. Which of the following represents the volume of blood passing through a vessel per unit of time?
  a. mean arterial pressure
  b. compliance
  c. flow rate
  d. distensibility
  e. vascular tone

_____ 74. Which of the following represents the average pressure responsible for driving blood forward into the tissue?
  a. negative thoracic pressure
  b. systolic pressure
  c. diastolic pressure
  d. mean arterial pressure
  e. none of the above

_____ 75. Which of the following represents the maximum pressure exerted in the arteries when blood is ejected into them during ventricular contraction?
  a. systolic pressure
  b. diastolic pressure
  c. venous pressure
  d. capillary pressure
  e. pulse pressure

_____ 76. Which of the following does not hasten the development of primary hypertension?
  a. obesity
  b. smoking
  c. stress
  d. excessive ingestion of salt
  e. none of the above

_____ 77. Which of the following represents the minimum pressure within the arteries when blood is draining off into the remaining vessels as the ventricles relax?
  a. means arterial pressure
  b. pulse pressure
  c. diastolic pressure
  d. venous pressure
  e. capillary pressure

_____ 78. Which of the following represents the difference between the maximum and minimum arterial pressures?
  a. mean arterial pressure
  b. pulse pressure
  c. diastolic pressure
  d. systolic pressure
  e. venous pressure

_____ 79. Which of the following represents a force caused by colloidal dispersion?
  a. CBP
  b. IFHP
  c. PCOP
  d. IHOP
  e. none of the above

_____ 80. Which of the following transport absorbed fat from the digestive system?
  a. veins
  b. arteries
  c. venules
  d. capillaries
  e. lymph vessels

_____ 81. Which of the following have little inherent tone and little elastic recoil?
  a. arteries
  b. veins
  c. capillaries
  d. lymphatics
  e. all the above

_____ 82. Which of the following imparts strength to the walls of arteries?
  a. striated muscles
  b. endothelin
  c. endothelium
  d. elastin
  e. collagen fibers

**Points to Ponder**
1. Having learned about active hyperemia can you think of an example of passive hyperemia?
2. How do the palace guards in London avoid fainting?
3. Why is the cardiovascular center in the medulla as opposed to some cortical, hypothalamic or thalamic area?

**Experiment of the Day**

1. On a classmate's arm try to locate the venous valves?

# BLOOD AND BODY DEFENSES  9

## CHAPTER OVERVIEW

Blood plays a vital role in homeostasis through the transport of nutrients, dissolved gases, wastes and substances such as hormones.  While blood and circulatory system are of no greater importance than many other systems or organs they are one of the weaker links in maintaining homeostasis.  There are two kidneys; things can be relearned; and the liver can regenerate, but breakdowns in the blood and circulatory system can result in greater consequences.

There are two general components in blood besides water:  plasma and the cellular elements.  All vascular transportation relies on the fluid portion of the blood.  In addition, the plasma maintains an osmotic gradient that is important in the distribution of extracellular fluid.  The plasma also has a buffer system against changes in pH and the necessary constituents to begin the clotting process.

Erythrocytes contain hemoglobin which plays a vital role in oxygen and carbon dioxide transport and the buffering capacity of the blood.  Anemia results any time the oxygen carrying capacity is reduced.  There are many causes of anemia.

Leukocytes defend the body against pathogens, identify and destroy cancer cells and clean up cellular debris.  Platelets are the major elements in maintaining homeostasis.  All the cellular components of blood are produced in the bone marrow even though lymphocytes simply divide in the normal healthy body.

The immune system contributes to homeostasis by providing a sophisticated highly effective defense system.  The system is essential in defending against viruses and bacteria,  removing "worn out" cell and tissue debris, and destroying mutant cells.  However, these benefits are not without cost.  Inappropriate immune responses and tissue rejection are problems due to the efficacy of the immune system as a defense system.

The primary defensive cells are neutrophils, eosinophils, basophils, B lymphocytes, T lymphocytes and macrophages.

The methods of defense include phagocytizing, lysing secreting chemicals, marking for destruction, producing antibodies and inactivating the invaders. Also involved are memory and the transfer of information by cloning. The immunity derived from this system is either active, passive, or acquired. In most systems the cells function as part of a tissue or organ while in the immune system the defense is primarily from individual cells.

The skin and all surfaces that communicate with the external environment provide either physical and/or chemical defenses. Defense of the body begins at the environmental-body interface and continues throughout the body. Thus, homeostasis is maintained for the survival of cells, which make up body systems, which maintain homeostasis.

# CHAPTER OUTLINE

## Introduction (Text page 277)

Blood consists of three types of specialized cellular elements, erythrocytes, leukocytes and platelets, suspended in the complex liquid plasma.

The hematocrit represents the percentage of total blood volume occupied by erythrocytes. Plasma accounts for the remaining volume.

The white blood cells and platelets represent less than one percent of the blood volume.

## Plasma (Text page 278)

Many of the functions of plasma are carried out by plasma proteins (page 278).

Plasma is composed of ninety percent water.

Because water has a high capacity to hold heat, plasma is able to absorb and distribute much of the heat generated metabolically within the tissues.

Heat energy not needed to maintain body temperature is eliminated to the environment.

The most plentiful organic constituents by weight are plasma proteins, which compose six percent to eight percent of plasma total weight.

Inorganic constituents account for approximately one percent of plasma weight.

The most abundant electrolytes in the plasma are sodium and chloride ions.

The remaining small percentage of plasma is occupied by nutrients, waste products, dissolved gases and hormones.

Most of these substances are merely being transported in the plasma.

The plasma proteins are one group of plasma constituents not present just for the ride.

Because they are the largest of the plasma constituents, plasma proteins usually do not exit through the narrow pores in the capillary walls.

Plasma proteins exist in a colloidal dispersion.

There are three groups of plasma proteins--albumins, globulins, and fibrinogen.

Plasma proteins establish an osmotic gradient between blood and interstitial fluid.

Plasma proteins are partially responsible for the plasma's capacity to buffer changes in pH.

Plasma proteins contribute to blood viscosity.

Plasma proteins can be degraded to provide energy for cells.

Plasma proteins bind many substances for transport through the plasma.

Alpha and beta globulins are involved in the process of blood clotting.

Inactive precursor protein molecules, which are activated as needed by specific regulatory inputs, are plasma proteins.

The gamma globulins are the immunoglobulins.

Plasma proteins are key factors in the blood-clotting process.

# Erythrocytes (Text page 278)

The structure of erythrocytes is well suited to their primary function of oxygen transport in the blood (page 278).

Erythrocytes are flat biconcave discs.

The biconcave shape provides a larger surface area for diffusion of oxygen.

The thinness of the cell enables oxygen to diffuse rapidly between the exterior and the innermost regions of the cell.

Erythrocytes most important feature that enables them to transport oxygen is the hemoglobin they contain.

Hemoglobin consists of (1) the globin portion, a protein made up of four highly folded polypeptide chains, and (2) four iron-containing heme groups.

Each hemoglobin molecule can pick up four oxygen passengers.

Hemoglobin can also combine with carbon dioxide, the acidic hydrogen-ion portion of ionized carbonic acid, and carbon monoxide.

Hemoglobin plays the key role in oxygen transport while contributing significantly to carbon dioxide transport and the buffering capacity of blood.

Erythrocytes contain no nucleus, organelles, or ribosomes.

The bone marrow continuously replaces worn-out erythrocytes (page 279).

Most old red blood cells meet their final demise in the spleen.

The spleen has a limited ability to store healthy erythrocytes in its pulpy interior; serves as a reservoir site for platelets; and contains an abundance of lymphocytes.

Old erythrocytes must be replaced by new cells produced in the bone marrow.

The bone marrow normally generates new red blood cells, a process known as erythropoiesis, keeping pace with the demolition of old cells.

Red bone marrow that is capable of blood-cell production is found in the sternum, vertebrae, ribs, base of the skull, and upper ends of the long limb bones.

Red marrow not only produces red blood cells but is the ultimate source for leukocytes and platelets.

Undifferentiated pluripotential stem cells continuously divide and differentiate to give rise to each of the types of blood cells.

Regulatory factors act on the hemopoietic marrow to govern the number and type of cells generated and discharged into the blood.

Erythropoiesis is controlled by erythropoietin from the kidneys (page 280).

Reduced oxygen delivery to the kidneys stimulates them to secrete the hormone erythropoietin into the blood.

This hormone stimulates erythropoiesis by the bone marrow.

Anemia can be caused by a variety of disorders (page 280).

Anemia refers to a reduction below normal in the oxygen-carrying capacity of the blood.

Nutritional anemia is caused by a dietary deficiency of a factor needed for erythropoiesis.

Iron-deficiency anemia occurs when insufficient iron is available for the synthesis of hemoglobin.

Pernicious anemia is caused by an inability to absorb adequate amounts of vitamin $B_{12}$ from the digestive tract.

Aplastic anemia is caused by failure of the bone marrow to produce adequate numbers of red blood cells.

Renal anemia is caused by inadequate erythropoietin secretion as a result of kidney disease.

Hemorrhagic anemia is caused by the loss of substantial quantities of blood.

Hemolytic anemia is caused by the rupture of excessive numbers of circulating erythrocytes.

## Polycythemia is an excess of circulating erythrocytes (page 282).

Polycythemia is characterized by an excess of circulating red blood cells.

Primary polycythemia is caused by a tumorlike condition of the bone marrow in which the erythropoiesis proceeds at an excessive uncontrolled rate.

Secondary polycythemia is an appropriate erythropoietin-induced adaptive mechanisms to improve the blood's oxygen-carrying capacity.

# Platelets and Hemostasis (Text page 282)

## Platelets are cell fragments derived from megakaryocytes (page 282).

Platelets are another type of cellular element present in the blood.

Platelets are small cell fragments that have budded off the outer edges of extraordinarily large bone-marrow-bound cells known as megakaryocytes.

## Hemostasis prevents blood loss from small damaged vessels (page 282).

Hemostasis is the arrest of bleeding from a broken blood vessel.

Hemostasis involves three major steps: (1) vascular spasm, (2) formation of a platelet plug, and (3) blood coagulation.

## Vascular spasm reduces blood flow through an injured vessel (page 283).

A cut or torn blood vessel immediately constricts.

The opposing endothelial surfaces of the vessel are pressed together by this initial vascular spasm.

The surfaces become sticky and adhere to each other.

## Platelets aggregate to form a plug at a vessel defect (page 283).

When the smooth endothelial lining is disrupted because of vessel injury, platelets attach to the exposed collagen.

Platelets release adenosine diphosphate which causes the surface of nearby circulating platelets to become sticky so that they adhere to the first layer of platelets.

A plug of platelets is rapidly built up at the defect site.

## A triggered chain reaction involving clotting factors in the plasma results in blood coagulation (page 284).

Blood coagulation or clotting is the transformation of blood from a liquid into a solid gel.

The conversion of fibrinogen into fibrin is catalyzed by the enzyme thrombin at the site of vessel injury.

Fibrin molecules adhere to the damaged vessel surface, forming a loose, netlike meshwork, or clot, that traps the cellular elements of the blood.

Thrombin exists in the plasma in the form of an inactive precursor called prothrombin. Factor X converts prothrombin into thrombin.

Altogether, twelve plasma clotting factors participate in essential steps that lead to the final conversion of fibrogen into a stabilized fibrin network.

Once the first factor in the sequence is activated, it in turn activates the next factor, and so on, in a series of sequential reactions known as a cascade, until thrombin catalyzes the final conversion of fibrinogen into fibrin.

The clotting cascade may be triggered by the intrinsic pathway or extrinsic pathway. The intrinsic pathway is set off when factor XII is activated by coming in contact with exposed collagen.

The extrinsic pathway requires contact with tissue factors external to the blood.

When a tissue is traumatized, it releases a protein complex known as tissue thromboplastin which directly activates factor X, thereby bypassing all preceding steps of the intrinsic pathway.

## Fibrinolytic plasmin dissolves clots and prevents inappropriate clot formation (page 286).

Fibroblasts form a scar at the vessel defect.

The clot is slowly dissolved by a fibrinolytic enzyme called plasmin.

Phagocytic white blood cells gradually remove the products of clot dissolution.

## Inappropriate clotting is responsible for thromboembolism (page 286).

An abnormal intravascular clot attached to a vessel wall is known as a thrombus, and free floating clots are called emboli.

Several factors can cause thromboembolism; (1) roughened vessel surfaces associated with atherosclerosis, (2) imbalances in the clotting-anticlotting systems, (3) slow-moving blood, and (4) the release of tissue thromboplastin.

## Hemophilia is the primary condition responsible for excessive bleeding (page 286).

The most common cause of excessive bleeding is hemophilia, which is caused by a deficiency in one of the factors in the clotting cascade.

Eighty percent of all hemophiliacs lack the genetic ability to synthesize factor VIII.

Vitamin K deficiency can cause a bleeding tendency.

# Leukocytes (Text page 287)

## Leukocytes function primarily outside the blood (page 287).

Leukocytes are the mobile units of the body's immune defense system.

The leukocytes and their derivatives (1) defend against invasion by pathogens by phagocytizing the foreigners or causing their destruction by more subtle means; (2) identify and destroy cancer cells that arise within the body; and (3) function as a "cleanup crew" that removes the body's "litter" by phagocytizing debris resulting from dead or injured cells.

Pathogenic bacteria and viruses are the major targets of the immune defense system (page 287).

> Pathogenic bacteria that invade the body induce tissue damage and produce disease largely by releasing enzymes or toxins that physically injure or functionally disrupt affected cells and organs.
>
> Because viruses lack cellular machinery for energy production and protein synthesis, they are unable to carry out metabolism and reproduce unless they invade a host cell.
>
> Most viruses lead to damage or death of the cells they invade.

There are five different types of leukocytes (page 287).

> The five types of leukocytes fall into two main categories; (1) polymorphonuclear agranulocytes (neutrophils, eosinophils, and basophils) and (2) mononuclear agranulocytes (monocytes and lymphocytes).

Leukocytes are produced at varying rates, depending on the changing defense needs of the body (page 288).

> Granulocytes and monocytes are produced only in the bone marrow.
>
> Lymphocytes are actually produced by already-existing lymphocytes residing in the lymphoid tissues.
>
> Neutrophils are phagocytic specialists.
>
> An increase in circulating eosinophils is associated with allergic conditions and internal parasite infestations such as worms.
>
> Eosinophils attach to the worm and secrete substances that kill it.
>
> Basophils synthesize and store histamine and heparin.
>
> Histamine release is important in allergic reactions, whereas heparin hastens the removal of fat particles from the blood following a fatty meal.
>
> Monocytes are destined to become professional phagocytes.
>
> Monocytes mature and enlarge to become the large tissue phagocytes known as macrophages.
>
> There are two types of lymphocytes, B lymphocytes and T lymphocytes.
>
> B lymphocytes produce antibodies which circulate in the blood.
>
> An antibody binds with and marks for destruction specific kinds of foreign matter.
>
> T lymphocytes destroy their specific target cells, a process known as cell-mediated immune response.

# Nonspecific Immune Responses (page 290)

Immune responses may be either nonspecific or specific (page 290).

> Nonspecific immune responses are inherent defense responses that nonselectively defend against foreign or abnormal material of any type.
>
> Specific immune responses are selectively targeted against particular foreign material to which the body has previously been exposed.

Nonspecific defenses include inflammation, interferon, natural killer cells, and the complement system (page 290).

> Nonspecific defenses that come into play whether or not there has been prior experience with the offending agent include the following:  inflammation, interferon, natural killer cells, and the complement system.

Inflammation is a nonspecific response to foreign invasion or tissue damage (page 290).

The ultimate goal of inflammation is to bring to the invaded or injured area phagocytes and plasma proteins that can (1) isolate, destroy, or inactivate the invaders; (2) remove debris; and (3) prepare for subsequent healing and repair.

Upon bacterial invasion macrophages already present in the area immediately begin phagocytizing the foreign microbes.

Almost immediately upon microbial invasion, arterioles within the area dilate, increasing blood flow to the site of injury.

Histamine is released in the area of tissue damage by mast cells.

Released histamine increases the capillaries permeability.

Plasma proteins that normally are prevented from leaving the blood escape into the inflamed tissue.

As the leaked plasma proteins accumulate in the interstitial fluid, they exert a colloid osmotic pressure.

The elevation in local osmotic pressure, and the increased capillary blood pressure result in localized edema.

Upon exposure to tissue thromboplastin in the injured tissue and to specific chemicals secreted by phagocytes on the scene, fibrinogen, the final factor in the clotting system, is converted into fibrin.

Fibrin forms interstitial-fluid clots in the spaces around the bacterial invaders and damaged cells.

This walling off of the injured region from the surrounding tissues prevents or at least delays the spread of bacterial invaders and their toxic products.

Within an hour after the injury, the area is teeming with leukocytes that have exited from the vessels.

Neutrophils are the first to arrive.

During the next eight to twelve hours, monocytes swell and mature into macrophages.

Leukocyte emigration from the blood into the tissues involves the processes of amoeboid movement and chemotaxins.

Within a few hours after the onset of the inflammatory response, the number of neutrophils in the blood may increase up to four or five times that of normal.

Dead tissue and many foreign materials have surface characteristics that differ from normal body cells.

Foreign particles are deliberately marked for phagocytic ingestion by being coated with chemical mediators (opsonins) generated by the immune system.

Microbe-stimulated phagocytes release many chemicals, which function as mediators of the inflammatory response.

Phagocytic secretions stimulate the release of histamine from mast cells.

Phagocytic secretions trigger both clotting and anticlotting systems.

Phagocytic secretions induce the development of fever by the release of endogenous pyrogen.

The ultimate purpose of the inflammatory process is to isolate and destroy injurious agents and to clear the area for tissue repair.

Salicylates and glucocorticoid drugs suppress the inflammatory response (page 293).

> Numerous drugs can suppress the inflammatory response; the most effective are the salicylates and glucocorticoids.
> Salicylates reduce fever by inhibiting the production of prostaglandins.
> Glucocorticoids suppress almost every aspect of the inflammatory response.

Interferon transiently inhibits multiplication of viruses in most cells (page 293).

> A nonspecific defense mechanism is the release of interferon from virus-infected cells.
> Interferon binds with receptors on the plasma membranes of neighboring cells signaling these cells to prepare for the possibility of impending viral attack.
> Interferon triggers the production of viral-blocking enzymes by potential host cells.
> Interferon markedly enhances the actions of cell-killing cells.
> Interferon also slows cell division and suppresses tumor growth.

Natural killer cells destroy virus-infected cells and cancer cells upon first exposure to them (page 293).

> Natural killer cells are naturally occurring, lymphocyte-like cells that nonspecifically destroy virus-infected cells and cancer cells by directly lysing their membranes.

The complement system kills microorganisms directly on its own and in conjunction with antibodies and also augments the inflammatory response (page 294).

> The system derives its name from the fact that it complements the action of antibodies.
> Complement-induced lysis is the major means of directly killing microbes without phagocytizing them.
> Complement components augment the inflammatory response by serving as chemotaxins acting as opsonins.

## Specific Immune Responses:  General Concepts (Text page 295)

Specific immune responses include antibody-mediated immunity accomplished by B lymphocyte derivatives and cell-mediated immunity accomplished by T lymphocytes (page 295).

> There are two classes of specific immune responses:  antibody-mediated, or humoral, immunity involving the production of antibodies by B lymphocytes derivatives known as plasma cells and cell-mediated immunity involving the production of activated T lymphocytes, which directly attack unwanted cells.
> During fetal life and early childhood, some of the immature lymphocytes migrate through blood to the thymus, where they undergo further processing to become T lymphocytes.
> Lymphocytes that mature without benefit of "thymic education" become B lymphocytes.
> The thymus gradually atrophies and becomes less important as the individual matures.
> The thymus does, however, continue to produce thymosin, a hormone important in maintaining the T cell lineage.

An antigen induces an immune response against itself (page 296).

> Both B and T cells must be able to specifically recognize unwanted cells and other material to be destroyed or neutralized as being distinct from the body's own normal cells.
>
> The presence of antigens enables them to make this distinction.
>
> Foreign proteins are the most common antigens.
>
> Haptens are low-molecular weight organic substances that are not antigens by themselves but can become antigens if they attach to body proteins.

# B Lymphocytes:  Antibody-Mediated Immunity (Text page 296)

Antibodies amplify the inflammatory response to promote destruction of the antigen that stimulated their production (page 296).

> In the case of B cells, binding with an antigen induces the cell to differentiate into a plasma cell, which produces antibodies that are able to combine with the specific antigen that stimulated the antibodies' production.
>
> Antibodies gain access to the blood or lymph where they are known as gamma globulins or immunoglobulins.
>
> Antibodies are grouped into five subclasses.
>
> Antibody proteins in all five subclasses are composed of four interlinked polypeptide chains--arranged in the shape of a Y.
>
> Characteristics of the arm regions of the Y determine the specificity of the antibody.
>
> An antibody has two identical antigen-binding sites, one at the tip of each arm.
>
> These antigen-binding fragments (FAB) are unique for each different antibody.
>
> The tail portion of every antibody within each immunoglobulin subclass is identical.
>
> Immunoglobulins cannot directly destroy foreign organisms or other unwanted materials upon binding with antigens on their surfaces.
>
> Antibodies mark or identify foreign material as targets for actual destruction by the complement system, phagocytes, or killer cells.

Each antigen stimulates a different clone of B lymphocytes to produce antibodies (page 299).

> Each B lymphocyte is preprogrammed to respond to only one of the millions of different antigens.
>
> The clonal selection theory proposes that diverse B lymphocytes are produced during fetal development, each capable of synthesizing an antibody against a particular antigen before ever being exposed to it.
>
> B cells remain dormant, not actually secreting their particular antibody product until they come in contact with the appropriate antigen.
>
> Antigen binding causes the activated B-cell clone to multiply and differentiate into two cell types--plasma cells and memory cells.
>
> In the blood, the secreted antibodies combine with invading free antigen, marking it for destruction by the complement system, phagocytic ingestion, or other means.
>
> A small proportion of the new B lymphocytes become memory cells, which do not participate in the current immune attack against the antigen but instead remain dormant and expand the specific clone.

Natural immunity is actually a special case of actively acquired immunity (page 301).

> Antibodies associated with blood types are the classic example of "natural antibodies". It is now known that individuals are unknowingly exposed at an early age to small amounts of A-and B-like antigens associated with common intestinal bacteria.
> No naturally occurring antibodies develop against the Rh factor.

Lymphocytes respond only to antigens that have been processed and presented to them by macrophages (page 303).

> Invading organisms or other antigens are first engulfed by macrophages.
> During phagocytosis, the macrophage processes the raw antigen intracellularly and then exposes the processed antigen on the outer surface of the macrophage's plasma membrane in such a way that the adjacent B cells can recognize and be activated by it.
> Macrophages secrete interleukin 1 that enhances the differentiation and proliferation of the now-activated B-cell clone.
> Interleukin 1 is also largely responsible for the fever and malaise accompanying many infections.

## T Lymphocytes:  Cell-Mediated Immunity (Text page 303)

The three types of T cells are specialized to kill virus-infected host cells and to help or suppress other immune cells (page 303).

> Unlike B cells, T cells do not secrete antibodies.
> T cells must be in direct contact with their targets, a process known as cell-mediated immunity.
> Like B cells, T cells are clonal and exquisitely antigen-specific.
> T cells are activated by foreign antigen only when it is present on the surface of a cell that also carries a marker of the individual's own identity.
> There are three subpopulations of T cells, depending on their roles when activated by antigen:  (1) cytotoxic T cells, (2) helper T cells, and (3) suppressor T cells.
> Like B cells, not all activated T-cell progeny become effector T cells.
> A small proportion of them remain dormant, serving as a pool of memory T cells that are primed and ready to respond should the same foreign antigen ever reappear within a body cell.
> The targets of cytotoxic T cells most frequently are host cells infected with viruses.
> One means by which cytotoxic T cells and natural killer cells destroy a targeted cell is by releasing perforin molecules, which penetrate into the target cell's surface membrane and join together to form porelike channels.
> The virus released upon destruction of the host cell is then directly destroyed in the extracellular fluid by phagocytic cells, neutralizing antibodies, and the complement system.
> Helper T cells secrete B-cell growth factor, which enhances the antibody-secreting ability of the activated B-cell clone.
> Helper T cells secrete T-cell growth factor, which augments the activity of cytotoxic T cells, suppressor T cells, and even other helper T cells responsive to the invading antigen.
> Some chemicals secreted by T cells act as chemotaxins to lure more neutrophils and macrophages-to-be to the invaded area.

Macrophage-migration inhibition factor, another important lymphokine released from helper T cells, keeps these large phagocytic cells in the region by inhibiting their outward migration.

The AIDS virus selectively invades helper T cells, destroying or incapacitating the cells that normally orchestrate much of the immune response.

The AIDS virus also invades macrophages.

Suppressor T cells limit the responses of all other immune cells.

### The immune system is usually tolerant of self-antigens (page 308).

Suppressor T cells probably also play an important role in preventing the immune system from attacking the person's own tissues, a phenomenon known as tolerance.

At least three different mechanisms appear to be involved in tolerance: (1) clonal deletion; (2) clonal anergy; and (3) inhibition by suppressor T cells.

A condition in which the immune system fails to recognize and tolerate self-antigens associated with particular tissues is known as an autoimmune disease, of which myasthenia gravis is an example.

### The major histocompatibility complex is the code for surface membrane-bound human leukocyte-associated antigens unique for each individual (page 308).

Self-antigens are plasma-membrane-bound glycoproteins, known as human leukocyte-associated antigens or HLA antigens.

T cells typically bind with HLA self-antigens only when they are in association with a foreign antigen.

T cells do bind with HLA antigens present on the surface of transplanted cells.

The ensuing destruction of the transplanted cells is responsible for rejection of transplanted or grafted tissues.

### Immune surveillance against cancer cells involves an interplay among cytotoxic T cells, natural killer cells, macrophages and interferon (page 309).

Another important function generally attributed to the T cell system is its role in recognizing and destroying newly arisen, potentially cancerous tumor cells before they have a chance to multiply and spread; a process known as immune surveillance.

Any normal cell may be transformed into a cancer cell if mutations occur within its genes responsible for controlling cell division.

Cancer cells typically remain immature and do not become specialized, often resembling embryonic cells instead.

Such dedifferentiated malignant cells lack the ability to perform the specialized functions of the normal cell type from which they mutated.

Potentially cancerous cells that do arise are usually destroyed by the immune system early in their development.

### A regulatory loop appears to link the immune system and the nervous and endocrine systems (page 310).

There are important links between the immune system and the body's two major control systems, the nervous and endocrine systems.

The immune system both influences and is influenced by the nervous and endocrine systems.

## Immune Diseases (Text page 311)

### Immune deficiency diseases reduce resistance to foreign invaders (page 311).

Abnormal functioning of the immune system can lead to immune diseases in two general ways: deficiency diseases and inappropriate immune attacks.

Deficiency diseases occur when the immune system fails to respond adequately to foreign invasion.

The most recent and tragically the most common acquired immune deficiency disease is AIDS, which is caused by HIV, a virus that invades and incapacitates the critical helper T cells.

### Inappropriate immune attacks against harmless environmental substances are responsible for allergies (page 311).

The other category of immune diseases involves inappropriate specific immune attacks that cause reactions harmful to the body.

These include: (1) autoimmune responses, in which the immune system turns against one of the body's own tissues; (2) immune-complex disease, which involve overexuberant antibody responses that "spill over" and damage normal tissues; and (3) allergies.

An allergy is the acquisition of an inappropriate specific immune reactivity to a normally harmless environmental substance.

In immediate hypersensitivity, the allergic response appears within about twenty minutes after the sensitized individual is exposed to an allergen, whereas in delayed hyper-sensitivity, the reaction is not generally manifested until a day or so following exposure.

The most common allergens that provoke immediate hypersensitivities are pollen grains, bee stings, penicillin, certain foods, molds, dust, feathers and animal fur.

The following are among the most important chemicals released during immediate allergic reactions: (1) histamine; (2) slow-reactive substance of anaphylaxis; and (3) eosinophil chemotactic factor.

When large amounts of these chemical mediators gain access to the blood, the extremely serious systemic reaction known as anaphylactic shock occurs, which can lead to circulatory failure.

## External Defenses (Text page 313)

### The skin consists of an outer protective epidermis and an inner connective tissue dermis (page 313).

The epidermis consists of numerous layers of epithelial cells.

Epidermal cells are tightly bound together by spot desmosomes, which interconnect with intracellular keratin filaments to form a strong cohesive covering.

The dermis is a connective-tissue layer that contains many elastin fibers and collagen fibers, as well as an abundance of blood vessels and specialized nerve endings.

Special infoldings of the epidermis into the underlying dermis form the skin's exocrine glands--the sweat glands and the sebaceous glands--as well as the hair follicles.

### Specialized cells in the epidermis produce keratin and melanin and participate in immune defense (page 315).

Melanocytes produce the brown pigment melanin

Keratinocytes are specialists in keratin production, as they die, keratinocytes form the outer protective keratinized layer and are also responsible for generating hair and nails.

Langerhans cells present antigen to helper T cells.

Granstein cells interact with suppressor T cells, probably serving as a "brake" or skin-activated immune responses.

## Protective measures within body cavities that communicate with the external environment discourage pathogen invasion into the body (page 315).

Saliva secreted into the mouth at the entrance of the digestive system contains an enzyme that lyses certain bacteria.

Many of the surviving bacteria that are swallowed are killed by the strongly acidic gastric juice that they encounter in the stomach.

Some bacteria do manage to survive and reach the large intestine.

These harmless resident flora competitively suppress the growth of potential pathogens that have escaped the antimicrobial measures of earlier parts of the digestive tract.

Within the genitourinary system, would-be invaders encounter hostile conditions in the acidic urine and acidic vaginal secretions.

The genitourinary organs also produce a sticky mucus, which, like flypaper, entraps small invading particles.

Large airborne particles are filtered out of the inspired air by hairs at the entrances of the nasal passages.

The respiratory airways are coated with a layer of thick, sticky mucus secreted by epithelial cells within the airway lining.

This mucus sheet, laden with any inspired particulate debris is constantly moved upward to the throat by ciliary action.

Also contributing to defense against respiratory infections are antibodies secreted in the mucus.

In addition, an abundance of phagocytic specialists called the alveolar macrophages scavenge within the air sacs of the lungs.

Cigarette smoking suppresses these normal respiratory defenses.

## KEY TERMS

Agglutination-p297
Allergen-p311
Allergy-p311
Antibodies-p289
Antigen-p296
Autoimmune disease-p308
Basophils-p288
Cell-mediated immunity-p295
Complement system-p294
Emboli-p286
Endogenous pyrogen-p292
Eosinophils-p288
Erythropoiesis-p280
Fibrin-p284
Fibrinogen-p284
Gamma globulins-p296
Granstein cells-p315

Haptens-p296
Hematocrit-p278
Hemoglobin-p278
Hemophilia-p286
Humoral immunity-p295
Interferon-p293
Interleukin 1-p303
Interleukin 2-p306
Killer cells-p299
Langerhans cells-p315
Lymphokines-p306
Macrophages-p289
Megakaryocytes-p282
Memory cells-p299
Metastasis-p309
Monocytes-p288
Neutrophils-p288

Opsonins-p292
Pathogens-p287
Perforin-p306
Plasma cell-p296
Plasma proteins-p278
Plasmin-p286
Plasminogen-p286
Pluripotential stem cells-p280
Prothrombin-p284
Serum-p286
Spleen-p279
Thrombin-p284
Thrombus-p286
Thymosin-p295
Tissue thromboplastin-p285
Vascular spasm-p283

## REVIEW EXERCISES

**True - False** (Answers on page A-20)

_____ 1. Erythrocytes most important feature that enables them to transport oxygen is the hemoglobin they contain.

_____ 2. Hemoglobin can also combine with carbon monoxide.

_____ 3. Each of us has a total of 25 to 30 billion red blood cells.

_____ 4. Red marrow not only produces red blood cells but it is the ultimate source for leukocytes and platelets as well.

_____ 5. Nutritional anemia is caused by an inability to absorb adequate amounts of vitamin $B_{12}$ from the digestive tract.

_____ 6. Secondary polycythemia is caused by a tumorlike condition of the bone marrow in which erythropoiesis proceeds at an excessive, uncontrolled rate.

_____ 7. The conversion into fibrin is catalyzed by the enzyme thromboxane at the site of vessel injury.

_____ 8. The clotting cascade may be triggered by the intrinsic pathway or the extrinsic pathway.

_____ 9. Intrinsic pathway requires only four steps to bring about clotting.

_____ 10. Phagocytic white blood cells gradually remove the products of clot dissolution.

_____ 11. An abnormal intravascular clot attached to a vessel wall is called a thrombus.

_____ 12. Eighty percent of all hemophiliacs lack the genetic ability to synthesize factor VIII.

_____ 13. Vitamin A deficiency can cause a bleeding tendency.

_____ 14. Plasma is composed of eighty percent water.

_____ 15. Inorganic constituents account for approximately one percent of plasma weight.

_____ 16. The most abundant electrolytes in the plasma are sodium and potassium.

_____ 17. Plasma proteins usually do not exit through the narrow pores in the capillary walls.

_____ 18. Plasma proteins exist in a colloidal dispersion.

_____ 19. Plasma proteins contribute to blood viscosity.

_____ 20. The gamma globulins are the immunoglobulins.

_____ 21. Plasma proteins do not affect the plasma's capacity to buffer changes in pH.

_____ 22. Leukocytes lack hemoglobin.

_____ 23. Eosinophils have an affinity for the blue dye.

_____ 24. Monocytes have a large spherical nucleus that occupies most of the cell.

_____ 25. Granulocytes and monocytes are produced only in the bone marrow.

_____ 26. The total number of leukocytes normally ranges from five to ten million cells per milliliter of blood.

_____ 27. Basophils store histamine and heparin.

_____ 28. Heparin release is important in allergic reactions.

_____ 29. T lymphocytes produce antibodies.

_____ 30. Lymphocytes that mature without benefit of "thymic education" become T lymphocytes.

_____ 31. During fetal development some immature lymphocytes migrate to the thymus, where they undergo further processing to become T lymphocytes.

_____ 32. There are four classes of specific immune responses.

_____ 33. Foreign proteins are the most common antigens.

_____ 34. The thymus becomes less important as the individual matures.

_____ 35. Antibodies are secreted into the blood or lymph.

_____ 36. Antibodies are grouped into six subclasses.

_____ 37. Immunoglobulins cannot directly destroy foreign organisms.

_____ 38. Each B lymphocyte is preprogrammed to respond to only one of the millions of different antigens.

_____ 39. There are two naturally occurring antibodies developed against the Rh factor.

_____ 40. The most common allergens that provoke immediate hypersensitivities are pollen grains, bee stings, penicillin, certain foods, molds, dust, feathers, and animal fur.

_____ 41. Severe hypotension can lead to circulatory failure.

_____ 42. Immune complex disease is when the immune system turns against one of the body's own tissues.

_____ 43. In immediate hypersensitivity, the allergic response appears within about ten minutes after being exposed to an allergen.

_____ 44. T cells do not secrete antibodies.

_____ 45. Antibodies not only target viruses for destruction in the extracellular fluid but can also eliminate viruses inside neurons.

_____ 46. Helper T cells secrete B-cell growth factor.

_____ 47. The AIDS virus selectively invades cytotoxic T cells.

_____ 48. Self-antigens are plasma-membrane-bound glycoproteins.

_____ 49. B cells typically bind with HLA self-antigens only when they are in association with a foreign antigen.

_____ 50. Most body cells undergo mutations that result in malignancy.

_____ 51. Epidermal cells are tightly bound together by spot desmosomes.

_____ 52. The epidermis has very little direct blood supply.

_____ 53. The keratinized layer is airtight, fairly waterproof, and impervious to most substances.

_____ 54. Epidermal enzymes are not able to convert many potential carcinogens into harmless compounds.

_____ 55. The dermal blood vessels have no control in temperature regulation.

_____ 56. Langerhans cells present antigen to helper T cells.

_____ 57. T lymphocytes secrete antibodies that indirectly lead to the destruction of foreign material.

_____ 58. Natural killer cells is a family of proteins that nonspecifically defend against viral infections.

_____ 59. Released histamine increases the capillaries permeability.

_____ 60. Neutrophils are the first to arrive at the site of an injury.

_____ 61. Salicylates reduce fever by inhibiting the production of prostaglandins.

**Fill in the Blank** (Answers on page A-21)

62. _____ is the arrest of bleeding from a broken blood vessel.

63. _____ or _____ is the transformation of blood from a liquid into a solid gel.

64. When a tissue is traumatized, it releases a protein complex known as _____.

65. _____ is the fluid squeezed from the clot during clot retraction.

66. The clot is slowly dissolved by a fibrinolytic enzyme called _____.

67. Free floating clots are called _____.

68. The most common cause of excessive bleeding is _____.

69. Most old red blood cells meet their final demise in the _____.

70. The bone marrow normally generates new red blood cells, a process known as _____.

71. _____ anemia is caused by failure of the bone marrow to produce adequate numbers of red blood cells.

72. _____ refers to a reduction below normal in the oxygen carrying capacity of the blood and is characterized by a low hematocrit.

73. _____ anemia is caused by the rupture of excessive numbers of circulating erythrocytes.

74. Hemolysis occurs in defective cells such as in the disease _____.

75. The remaining small percentage of plasma is occupied by (1)_____, (2) _____, (3) _____, and (4) _____.

76. _____ are the one group of plasma constituents not present just for the ride.

77. Name the three groups of plasma proteins; (1) _____, (2) _____ and (3) _____.

78. _____ bind many substances for transport through the plasma.

79. _____ are the mobile units of the body's immune defense system.

80. Neutrophils, eosinophils, and basophils are categorized as _____.

81. _____ are the smallest of the leukocytes.

82. Leukocytes and their derivatives defend against invasion by _____.

83. _____ are phagocytic specialists.

84. _____ are the least numerous and most poorly understood of the leukocytes.

85. T lymphocytes destroy their specific target cells, a process known as a _____ _____.

86. _____ are destined to become professional phagocytes.

87. Antibodies are secreted into the blood or lymph, where they are known as _____ _____ or _____.

88. _____ is the term applied to the process in which foreign cells, such as bacteria or mismatched transfused red blood cells, bind together a clump.

89. Antibodies mark or identify foreign material as targets for actual destruction by the _____, _____, or _____.

90. _____ remain dormant, not actually secreting their particular antibody product until they come into contact with the appropriate _____.

91. The production of antibodies as a result of exposure to an antigen is referred to as _____ against that antigen.

92. Antigen-presenting macrophages secrete _____, a multipurpose chemical mediator that enhances the differentiation and proliferation of the now-activated B-cell clone.

93. T-cells must be in direct contact with their targets, a process known as _____ _____.

94. _____ destroy host cells bearing foreign antigen, such as body cells invaded by viruses, cancer cells, and transplanted cells.

95. Natural killer cells release _____ molecules, which penetrate into the target cell's surface membrane and join together to form porelike channels.

96. Helper T cells secrete T cell growth factor, also known as _____.

97. _____ are by far the most numerous of the T cells.

98. _____ is an example of an autoimmune disease.

99. The most recent and tragically the most common acquired immune deficiency disease is _____.

100. An _____ is the acquisition of an inappropriate specific immune reactivity, or _____, to a normally harmless environmental substance.

101. When large amounts of chemical mediators or allergens gain access to the blood, the extremely serious systemic reaction known as _____ occurs.

102. _____ refers to the body's ability to resist or eliminate potentially harmful foreign materials or abnormal cells.

103. _____ refer collectively to the tissues that store, produce, or process lymphocytes.

104. Localized vasodilation is primarily induced by _____ that has been released in the area of tissue damage from mast cells.

105. Upon exposure to tissue thromboplastin in the injured tissue and to specific chemicals secreted by phagocytes on the scene, _____, the final factor in the clotting system, is converted into _____.

106. Numerous drugs can suppress the inflammatory process; the most effective are the _____ and related compounds and _____.

107 There are two classes of specific immune responses: _____and

_____.

108. The _____ gradually atrophies and becomes less important as the individual matures. It does, however, continue to produce _____, a hormone important in maintaining the T-cell lineage.

109. _____ are low-molecular-weight organic substances that are not antigenic by themselves but can become antigenic if they attach to body proteins.

110. In _____, the allergic response appears within about twenty minutes after a sensitized individual is exposed to an allergen.

111. Epidermal cells are tightly bound together by _____.

112. The cells of the _____ produce an oily secretion known as

_____.

113. _____ release a dilute salt solution through small openings, _____, onto the surface of the body.

114. _____ produce the brown pigment _____.

115. The most abundant epidermal cells are the _____.

116. _____ present an antigen to helper T cells, thereby facilitating their responsiveness to skin-associated antigens.

117. _____ interact with suppressor T cells, probably serving as a "brake" on skin-activated immune responses.

118. Phagocytic specialists called the _____ scavenge within the air sacs of the lungs.

**Matching:** Match the blood constituent to their function (Answers on page A-22)

| | | |
|---|---|---|
| _____ 119. Carries heat | a. platelets |
| _____ 120. Hemostasis | b. erythrocyte |
| _____ 121. Become macrophages | c. fibrinogen |
| _____ 122. Transport oxygen | d. basophils |
| _____ 123. Precursor for fibrin | e. neutrophils |
| _____ 124. Cell-mediated immune response | f. eosinophils |
| _____ 125. Release histamine | g. B lymphocytes |
| _____ 126. Attacks parasitic worms | h. water |
| _____ 127. Antibody production | i. T lymphocytes |
| _____ 128. Phagocytizes bacteria | j. monocytes |

**Matching:** Match the lymphoid tissue to the function (Answers on page A-22)

_____ 129. Exchanges lymphocytes with blood
_____ 130. Origin of all blood cells
_____ 131. Stores a small percentage of red blood cells
_____ 132. Secretes the hormone thymosin
_____ 133. Exchanges lymphocytes with lymph
_____ 134. Site of maturational processing for T lymphocytes
_____ 135. Resident lymphocytes produce antibodies and sensitized T cells, which are released into the blood
_____ 136. Site of maturational processing for B lymphocytes
_____ 137. Resident macrophages remove microbes and other particulate debris from lymph
_____ 138. Resident lymphocytes produce antibodies and sensitized cells, which are released into the lymph
_____ 139. Resident macrophages remove microbes and other particulate debris, most notably worn-out red blood cells, from the blood

a. thymus
b. spleen
c. bone marrow
d. lymph nodes, tonsils, adenoids, appendix, gut-associated lymphoid tissue

**Multiple Choice** (Answers on page A-22)

_____ 140. Which cellular element has the largest percentage of total blood volume?
     a. eosinophils
     b. erythrocytes
     c. basophils
     d. neutrophils
     e. platelets

_____ 141. Which type of anemia is caused by failure of the bone marrow to produce adequate numbers of red blood cells?
     a. nutritional anemia
     b. pernicious anemia
     c. hemolytic anemia
     d. aplastic anemia
     e. renal anemia

_____ 142. Which type of anemia is caused by the loss of substantial quantities of blood?
     a. nutritional anemia
     b. hemorrhagic anemia
     c. aplastic anemia
     d. renal anemia
     e. pernicious anemia

_____ 143. Which of the plasma proteins are antibodies?
      a. fibrinogen
      b. alpha globulins
      c. gamma globulins
      d. beta globulins
      e. albumins

_____ 144. Which organ secretes erythropoietin?
      a. bone marrow
      b. heart
      c. lungs
      d. spleen
      e. kidney

_____ 145. Which type of anemia is caused by inadequate erythropoietin production?
      a. nutritional anemia
      b. pernicious anemia
      c. hemolytic anemia
      d. aplastic anemia
      e. renal anemia

_____ 146. Which disorder is characterized by a tumorlike condition of the bone marrow?
      a. hemorrhagic anemia
      b. sickle-cell anemia
      c. secondary polycythemia
      d. leukemia
      e. primary polycythemia

_____ 147. Which substance causes platelets to become sticky?
      a. adenosine diphosphate
      b. plasminogen
      c. fibrinogen
      d. tissue thromboplastin
      e. prothrombin

_____ 148. Which factor do most hemophiliacs lack the ability to synthesize?
      a. factor X
      b. factor VII
      c. factor XII
      d. factor VIII
      e. factor XIII

_____ 149. Which substance slowly dissolves the clot?
      a. thromboplastin
      b. plasmin
      c. carbonic anhydrase
      d. fibrinogen
      e. plasminogen

_____ 150. Which of the following substances causes an inappropriate immune response?
      a. perforins
      b. opsonins
      c. endogenous pyrogens
      d. Haptens
      e. allergens

_____ 151. Which of the following substances induces the development of fever?
      a. perforins
      b. opsonins
      c. endogenous pyrogens
      d. Haptens
      e. allergens

_____ 152. Which of the following substances is a secretory product of helper T cells?
      a. kallikrein
      b. interleukin 1
      c. Haptens
      d. interleukin 2
      e. allergen

_____ 153. Which of the following substances make bacteria move susceptible to phagocytosis?
      a. allergens
      b. opsonins
      c. interleukin 1
      d. interleukin 2
      e. Haptens

_____ 154. Which of the following substances are not antigenic by themselves but can become antigenic if they attach to body proteins?
      a. opsonins
      b. allergens
      c. perforins
      d. histamines
      e. Haptens

_____ 155. Which of the following substances is secreted by helper T cells.
      a. histamine
      b. B cell growth factor
      c. endogenous pyrogen
      d. perforins
      e. interferon

_____ 156. Which of the following substances is a secretory product of macrophages?
      a. interleukin 1
      b. interferon
      c. kallikrein
      d. interkeukin 2
      e. perforin

_____ 157. Which of the following substances brings about vasodilation and increased capillary permeability?
      a. interferon
      b. histamine
      c. perforin
      d. interleukin 2
      e. kallikrein

_____ 158. Which of the following substances triggers the production of viral-blocking enzymes by potential host cells?
      a. perforin
      b. interleukin 2
      c. interleukin 1
      d. ferrokinogen
      e. interferon

_____ 159. Which of the following substances is released by cytotoxic T cells and binds to a target cell's surface membrane?
      a. perforin
      b. interferon
      c. interleukin 1
      d. interleukin 2
      e. ferrokinogen

**Points to Ponder**
1. Why do you suppose that the clotting cascade is so long and complicated?
2. Over the years we have gained much knowledge about the composition of blood. Why then is it so difficult to make artificial blood?
3. What is the advantage in having erythrocytes with no nuclei, organelles or ribosomes and are short-lived as compared to longer-lived erythrocytes that are complete cells.

4. With the current social views in this country do you feel that the irradication of AIDS is possible?
5. In that part of the body defenses involving B cells and T cells, cellular memory plays a major role. When and how is this memory lost?
6. Do animal rights activists hinder our fight against viral diseases?

**Clinical Perspective**
1. When a person cuts their wrists to commit suicide does it make a difference whether or not they completely sever the vessels?
2. What is the significance of a "booster shot?"
3. This summer you will direct a biological field trip to the remote areas of the Chihuahuan Desert. The students and yourself will spend a month living in tents as you study the desert. With regard to this chapter what precautions should you take before you begin this trip?

# RESPIRATORY SYSTEM 10

## CHAPTER OVERVIEW

The homeostasis role of the respiratory system overlaps that of the circulatory system: supplying oxygen to the cells of the body and removing carbon dioxide, a byproduct of cellular metabolism. While these are not the only functions of the respiratory system they are surely the most important. The entire ventilation-transport process, from air outside the body to the gaseous exchange at the systemic cellular level, involves diffusion down gradients. For the most part they are pressure gradients.

The mechanics of ventilation rely heavily upon the physical properties of the alveolar wall and surface tension. Ventilation is controlled by the nervous system at both conscious and subconscious levels.

The bonding of oxygen and carbon dioxide to hemoglobin is important for transport by the blood. Oxygen bonds to the heme portion and carbon dioxide attaches to the globin fraction of the hemoglobin molecule. This is however only one step in the process of respiration.

From alveolar air to the mitochondria of the various cells involves many barriers through which oxygen and carbon dioxide must traverse. Crossing the numerous barriers oxygen and carbon dioxide diffuse down pressure gradients involving differences in their respective partial pressures. Using quite sophisticated control mechanisms the respiratory system plays a major role in maintaining homeostasis.

# CHAPTER OUTLINE

## Introduction (Text page 322)

### The respiratory system does not participate in all steps of respiration (page 322).

Internal or cellular respiration refers to the intracellular metabolic processes carried out within the mitochondria, which use oxygen and produce carbon dioxide during the derivation of energy from nutrient molecules.

External respiration refers to the entire sequence of events in the exchange of oxygen and carbon dioxide between the external environment and the cells of the body.

External respiration, the topic of this chapter, encompasses four steps: (1) Air is alternately moved in and out of the lungs so that exchange of air can occur between the atmosphere and the alveoli of the lungs; (2) Oxygen and carbon dioxide are exchanged between air in the alveoli and blood within the pulmonary capillaries by the process of diffusion; (3) Oxygen and carbon dioxide are transported by the blood between the lungs and the tissues; and (4) The exchange of oxygen and carbon dioxide takes place between the tissues and the blood by the process of diffusion across the systemic capillaries.

The respiratory system additionally performs the following nonrespiratory functions: (1) It provides a route for water loss and heat elimination; (2) It enhances venous return; (3) It enables speech, singing, and other vocalization; (4) It defends against inhaled foreign matter; (5) It removes, modifies, activates, or inactivates various materials passing through the pulmonary circulation; and (6) The nose serves as the organ of smell; (7) It contributes to the maintenance of normal acid-base balance.

### The respiratory airways conduct air between the atmosphere and alveoli (page 322).

The respiratory system includes the respiratory airways leading into the lungs, the lungs themselves, and the structures of the thorax involved in producing movement of air through the airways into and out of the lungs.

The airways include; (1) nasal passages; (2) pharynx; (3) larynx; (4) trachea; (5) bronchi; and (6) bronchioles.

### The gas-exchanging alveoli are small, thin-walled, inflatable air sacs encircled by a jacket of pulmonary capillaries (page 323).

The alveoli are clusters of thin-walled, inflatable, grape-like sacs at the terminal branches of the conducting airways.

The alveolar walls consist of a single layer of flattened Type I alveolar cells.

The interstitial space between an alveolus and the surrounding capillary network forms an extremely thin barrier.

In addition to the Type I cells, the alveolar epithelium also contains Type II alveolar cells which secrete pulmonary surfactant.

### The lungs occupy much of the thoracic cavity (page 324).

There is no muscle within the alveolar walls to cause them to inflate or deflate during the breathing process.

The lungs, heart and associated vessels, esophagus, thymus and some nerves occupy the thoracic cavity.

The outer chest wall is formed by twelve pairs of curved ribs, which join the sternum anteriorly and the thoracic vertebrae posteriorly.
The diaphragm forms the floor of the thoracic cavity.

## A pleural sac separates each lung from the thoracic wall (page 325).

Separating each lung from the thoracic wall and other surrounding structures is a doubled-walled, closed sac called the pleural sac.
The surfaces of the pleura secrete a thin intrapleural fluid, which lubricates the pleural surfaces.

# Respiratory Mechanics (Text page 325)

## Interrelationships among atmospheric, intra-alveolar, and intrapleural pressures are important in respiratory mechanics (page 325).

Air tends to move from a region of higher pressure to a region of lower pressure down a pressure gradient.
Three different pressure considerations are important in ventilation; (1) atmospheric pressure, (2) intra-alveolar pressure, and (3) intrapleural pressure.

## The transmural pressure gradient holds the lungs and thoracic wall in tight apposition, even though the lungs are smaller than the thorax (page 327).

An even more important reason that the lungs follow the movements of the chest wall is the transmural pressure gradient that exists across the lung wall.
The intra-alveolar pressure is greater than the intrapleural pressure, so a greater pressure is pushing outward than is pushing inward across the lung wall.
If the intrapleural pressure were ever to equilibrate with the atmospheric pressure, the lungs and thorax would separate and assume their own inherent dimensions.

## Bulk flow of air into and out of the lungs occurs because of cyclical intra-alveolar pressure changes brought about indirectly by respiratory-muscle activity (page 328).

Because air flows down a pressure gradient, the intra-alveolar pressure must be less than atmospheric pressure for air to flow into the lungs during inspiration.
At the onset of inspiration, the inspiratory muscles--the diaphragm and external intercostal muscles--are stimulated to contract, resulting in enlargement of the thoracic cavity.
The intra-alveolar pressure is now less than atmospheric pressure, air flows into the lungs down the pressure gradient from higher to lower pressure.
During inspiration, the intrapleural pressure falls.
The resultant increase in the transmural pressure gradient during inspiration ensures that the lungs are stretched to fill the expanded thoracic cavity.
At the end of inspiration, the inspiratory muscles relax.
The chest wall and the stretched lungs recoil to their preinspiratory size because of their elastic properties.
In a resting expiration, the intra-alveolar pressure increases.
Air now leaves the lungs down a pressure gradient from high intra-alveolar pressure to lower atmospheric pressure.
Expiration is normally a passive process.

Inspiration is always an active process.

To produce active expiration, the expiratory muscles-the internal intercostal muscles and the muscles of the abdominal wall--must contract to further reduce the volume of the thoracic cavity and lungs.

## Airway resistance becomes an especially important determinant of airflow rates when the airways are narrowed by disease processes (page 331).

The primary determinant to resistance to airflow is the radius of the conducting airway.

Normally, modest adjustments in airway size can be accomplished by autonomic nervous system regulation to suit the body's needs.

Parasympathetic stimulation promotes bronchiolar smooth muscle contraction, which increases airway resistance by producing bronchoconstriction.

Sympathetic stimulation and its associated hormones, epinephrine, bring about bronchodilation and decreased airway resistance by promoting bronchiolar smooth-muscle relaxation.

Resistance becomes an extremely important impediment to airflow when airway lumens become narrowed as a result of disease.

Chronic obstructive pulmonary disease is a group of lung diseases characterized by increased airway resistance resulting from the narrowing of the lumen of the lower airways.

Chronic obstructive pulmonary disease encompasses three chronic diseases: asthma, chronic bronchitis, and emphysema.

When airway resistance is increased as a result of chronic obstructive lung disease of any type, expiration is more difficult to accomplish than inspiration.

## Elastic behavior of the lungs is due to elastic connective tissue fibers and alveolar surface tension (page 333).

Two interrelated concepts are involved in pulmonary elasticity: elastic recoil and compliance.

Elastic recoil refers to how readily the lungs rebound after having been stretched.

Compliance refers to how much effort is required to stretch or distend the lungs.

A highly compliant lung stretches further for a given increase in the pressure difference than does a less compliant lung.

Pulmonary elastic behavior depends mainly on two factors: highly elastic connective tissue in the lungs and alveolar surface tension.

Pulmonary connective tissue contains large quantities of elastin fibers arranged into a meshwork that amplifies their elastic behavior.

Surface tension is responsible for a twofold effect.

First, the liquid layer resists any force that increases its surface area.

Second, the liquid surface area tends to become as small as possible because the surface water molecules try to get as close together as possible.

## Pulmonary surfactant decreases surfaces tension and contributes to lung stability (page 335).

The tremendous surface tension of pure water is normally counteracted by secretion of pulmonary surfactant by the Type II alveolar cells.

By lowering the alveolar surface tension, pulmonary surfactant provides two important benefits: (1) it increases pulmonary compliance, thus reducing the work of inflating the lungs; and (2) it reduces the lungs' tendency to recoil so that they do not collapse readily.

Pulmonary surfactant's role in reducing the alveoli's tendency to recoil is important in helping maintain lung stability.

## A deficiency of pulmonary surfactant is responsible for newborn respiratory distress syndrome (page 335).

In an infant born prematurely, pulmonary surfactant may be insufficient to reduce the alveolar surface tension to manageable levels.

The resultant collection of symptoms that develop are referred to as newborn respiratory distress syndrome.

It is more difficult to expand a collapsed alveolus by a given volume than to increase an already partially expanded alveolus by the same volume.

With newborn respiratory distress syndrome, lung expansion may require considerably larger transmural pressure gradients.

The problem is compounded by the fact that the newborn's muscles are still weak.

## Normally, the lungs contain about 2 to 2.5 liters of air during the respiratory cycle but can be filled to over 5.5 liters or emptied to about 1 liter (page 336).

On average, the maximum amount of air that the lungs can hold is about 5.7 liters in males and 4.2 liters in females.

Anatomical build, age, the distensibility of the lungs, and the presence or absence of respiratory disease affect this total lung capacity.

At the end of a normal quiet expiration, the lungs still contain about 2,200 ml of air.

About 500 ml of air are inspired and the same quantity expired during quiet breathing.

Using a spirometer the following lung volumes and lung capacities can be determined: (1) tidal volume; (2) inspiratory reserve volume; (3) inspiratory capacity; (4) expiratory reserve volume; (5) residual volume; (6) functional residual capacity; (7) vital capacity; (8) total lung capacity; and (9) forced expiratory volume in one second.

Two general categories of respiratory dysfunction yield abnormal results during spirometry: obstructive and restrictive lung disease.

## Alveolar ventilation is less than pulmonary ventilation because of the presence of dead space (page 338).

Not all of the inspired air gets down to the site of gas exchange in the alveoli.

Part of the air remains in the conducting airways, where it is not available for gas exchange.

This volume is considered to be anatomic dead space.

Alveolar ventilation--the volume of air exchanged between the atmosphere and the alveoli per minute--is more important than pulmonary ventilation.

# Gas Exchange (Text page 339)

## Gases move down partial pressure gradients (page 339).

The ultimate purpose of breathing is to provide a continual supply of fresh oxygen for pickup by the blood and to constantly remove carbon dioxide unloaded from the blood.

Gaseous exchange at both the pulmonary-capillary and tissue-capillary level involves simple passive diffusion of oxygen and carbon dioxide down partial pressure gradients.
If, as is the case with oxygen, the alveolar partial pressure of a gas is higher than the partial pressure of that gas in the blood entering the pulmonary capillaries, the higher alveolar partial pressure drives the oxygen into the blood.
Conversely, if the alveolar partial pressure of a gas is lower than its partial pressure in the entering blood--the situation that exists for carbon dioxide--the lower alveolar partial pressure permits some of the carbon dioxide to escape from solution in the blood.
A gas always diffuses down its partial pressure gradient from the area of higher to the area of lower partial pressure.

## Oxygen enters and carbon dioxide leaves the blood in the lungs passively down partial pressure gradients (page 341).

The alveolar partial pressure of oxygen remains relatively constant throughout the respiratory cycle.
The partial pressure of oxygen in the blood likewise remains fairly constant at the same value.
A similar situation in reverse exists for carbon dioxide.
The appropriate partial pressure gradients between the alveoli and blood are maintained to ensure that oxygen enters the blood and carbon dioxide leaves the blood.
The amount of oxygen picked up in the lungs matches the amount extracted and used by the tissues.
The amount of carbon dioxide given up to the alveoli from the blood matches the amount of carbon dioxide picked up at the tissues.

## Factors other than the partial pressure gradient influence the rate of gas transfer (page 344).

According to Fick's law of diffusion, the rate of diffusion of a gas depends on the surface area and thickness of the membrane and on the diffusion coefficient of the gas.
Surface area is reduced in emphysema.
Inadequate gas exchange can also occur when the thickness of the barrier separating the air and blood is pathologically increased.
The thickness increases in (1) pulmonary edema, (2) pulmonary fibrosis, and (3) pneumonia.

## Gas exchange across the systemic capillaries also occurs down partial pressure gradient (page 344).

Just as they do at the pulmonary capillaries, oxygen and carbon dioxide move between the systemic capillary blood and the tissue cells by simple passive diffusion down partial pressure gradients.
The amount of oxygen transferred to the cells and the amount of carbon dioxide carried away from the cells depend on the rate of cellular metabolism.

# Gas Transport (Text page 345)

## Most oxygen in the blood is transported bound to hemoglobin (page 345).

Oxygen is present in the blood in two forms:  physically dissolved and chemically bound to hemoglobin.

Very little oxygen is physically dissolved in the plasma water because oxygen is poorly soluble in body fluids.

Of the oxygen transported by the blood, 98.5 percent is transported in combination with hemoglobin.

Hemoglobin, an iron-bearing protein molecule contained within the red blood cells, has the ability to form a loose easily reversible combination with oxygen.

## The partial pressure of oxygen is the primary factor determining the percent hemoglobin saturation (page 346).

Each hemoglobin molecule can carry up to four molecules of oxygen.

The most important factor determining the percent hemoglobin saturation is the partial pressure of oxygen of the blood, which in turn, is related to the concentration of oxygen physically dissolved in the blood.

When the partial pressure of oxygen in the blood is increased, as it is in the pulmonary capillaries, the reaction is driven toward an increase in the formation of oxyhemoglobin.

When the partial pressure of oxygen is decreased, as it is in the systemic capillaries, the reaction is driven toward an increase in the formation of reduced hemoglobin.

## By acting as a storage depot, hemoglobin promotes the net transfer of oxygen from the alveoli to the blood (page 347).

Net diffusion of oxygen from alveoli to blood occurs continuously until the hemoglobin becomes saturated with oxygen as completely as it can be at that particular partial pressure.

At a normal partial pressure for oxygen, hemoglobin is 97.5 percent saturated.

Not until hemoglobin can store no more oxygen does all of the oxygen transferred into the blood remain dissolved and directly contribute to the partial pressure.

Once the partial pressure of oxygen in the blood equilibrates with the alveolar partial pressure, no further oxygen transfer can take place.

The reverse situation occurs at the tissue level.

When the partial pressure of oxygen in the blood falls, hemoglobin is forced to unload some of its stored oxygen because the percent hemoglobin saturation is reduced.

Only when hemoglobin is no longer able to release any more oxygen into solution can the partial pressure of oxygen in the blood become as low as in the surrounding tissue.

At this time, further transfer of oxygen ceases.

Hemoglobin plays an important role in the total quantity of oxygen that the blood can pick up in the lungs and drop off in the tissues.

## Increased carbon dioxide, acidity, temperature, and 2,3-diphosphoglycerate shift the oxygen-hemoglobin dissociation curve to the right (page 348).

An increase in the parial pressure of carbon dioxide reduces the amount of oxygen and hemoglobin that can be combined.

An increase in acidity also reduces the amount of oxygen and hemoglobin that can be combined.

The influence of carbon dioxide and acid on the release of oxygen is known as the Bohr effect.

Local elevation in temperature enhances oxygen release from hemoglobin for use by the more active tissues.

Thus, increases in carbon dioxide, acidity, and temperature at the tissue level enhances the effect of a drop in the parial pressure of oxygen in facilitating the release of oxygen and hemoglobin.

These effects are largely reversed at the pulmonary level, where the extra acid-forming carbon dioxide is blown off and the local environment is cooler.

Inside the red blood cell, 2,3-diphosphoglycerate is produced during cellular metabolism. This erythrocyte constituent can bind reversibly with hemoglobin and reduce its affinity for oxygen.

## Oxygen-binding sites on hemoglobin have a much higher affinity for carbon monoxide than for oxygen (page 349).

Carbon monoxide and oxygen compete for the same binding sites on hemoglobin, but the affinity of hemoglobin for carbon monoxide is two hundred forty times that of the bond strength between hemoglobin and oxygen.

Fortunately, carbon monoxide is not a normal constituent of inspired air.

If carbon monoxide is being produced in a closed environment, it can reach lethal levels without the victim ever being aware of the danger.

## The majority of carbon dioxide is transported in the blood as bicarbonate (page 350).

Carbon dioxide is transported in the blood in three ways: (1) physically dissolved, (2) bound to hemoglobin, and (3) as bicarbonate.

The amount of carbon dioxide physically dissolved in the blood is only ten percent of the blood's total carbon dioxide content at normal systemic venous partial pressure levels.

Another thirty percent of the carbon dioxide combines with hemoglobin to form carbamino hemoglobin.

Carbon dioxide binds with the globin portion of hemoglobin.

By far the most important means of carbon dioxide transport is as bicarbonate.

Carbon dioxide combines with water to form carbonic acid.

This reaction proceeds swiftly within red blood cells because of the presence of the erythrocyte enzyme carbonic anhydrase which catalyzes the reaction.

The one carbon and two oxygen atoms of the original carbon dioxide molecule are thus present in the blood as an integral part of the bicarbonate ion.

Bicarbonate and hydrogen start to accumulate within the red blood cells in the systemic capillaries.

Reduced hemoglobin has a greater affinity for hydrogen ions than does oxyhemoglobin.

The Haldane effect and Bohr effect work in synchrony to facilitate oxygen liberation and the uptake of carbon dioxide and carbon dioxide-generated hydrogen ions.

## Various respiratory states are characterized by abnormal blood gas levels (page 351).

Hypoxia refers to insufficient oxygen at the cellular level.

There are four general categories of hypoxia: (1) hypoxic hypoxia, (2) anemic hypoxia, (3) circulatory hypoxia, and (4) histotoxic hypoxia.

Hypercapnia refers to excess carbon dioxide in the arterial blood.

Hypercapnia is caused by hypoventilation.

# Control of Respiration (Text page 353)

## Respiratory centers in the brain stem establish a rhythmic breathing pattern (page 353).

Respiratory control centers housed in the brain stem are responsible for generating the rhythmic pattern of breathing.

The primary respiratory control center is the medullary respiratory center.

There are two other respiratory centers higher in the brain stem in the pons--the apneustic center and the pneumotaxic center.

The dorsal respiratory group consists mostly of inspiratory neurons whose descending fibers terminate on the motor neurons that supply the inspiratory muscles.

The dorsal respiratory group is generally regarded as being responsible for the basic rhythm of ventilation.

The ventral respiratory group is called into play by the dorsal respiratory group as an "overdrive" mechanism during periods when demands for ventilation are increased.

The pontine centers exert "fine-tuning" influences over the medullary center to help produce normal, smooth inspirations and expirations.

The pneumotaxic center sends impulses to the dorsal respiratory group that help "switch off" the inspiratory neurons, thereby limiting the duration of inspiration.

The apneustic center prevents the inspiratory neurons from being switched off.

The pneumotaxic center is dominant over the apneustic center.

When the tidal volume is large, as during exercise, the Hering-Breuer reflex is triggered to prevent overinflation of the lungs.

Action potentials from the pulmonary stretch receptors travel through afferent nerve fibers to the medullary center and inhibit the inspiratory neurons.

## Carbon dioxide-generated hydrogen ion concentration in the brain extracellular fluid is normally the primary regulator of the magnitude of ventilation (page 354).

The medullary respiratory center receives inputs that provide information about the body's needs for gas exchange.

The arterial partial pressure of oxygen is monitored by peripheral chemoreceptors known as the carotid bodies and aortic bodies, which are located at the bifurcation of the common carotid arteries and in the arch of the aorta respectively.

The peripheral chemoreceptors are not sensitive to modest reductions in the arterial partial pressure of oxygen.

Reflex stimulation of respiration by the peripheral chemoreceptors serves as an important emergency mechanism in dangerously low arterial partial pressures of oxygen conditions.

The arterial partial pressure of carbon dioxide is the most important input regulating the magnitude of ventilation under resting conditions.

An increase in the arterial partial pressure of carbon dioxide reflexly stimulates the respiratory center, with the resultant increase in ventilation promoting elimination of excess carbon dioxide.

Conversely, a full in arterial partial pressure of carbon dioxide reflexly reduces respiratory drive.

There are no important receptors that monitor the arterial partial pressure of carbon dioxide per se.

Located in the medulla in the vicinity of the respiratory center, central chemoreceptors are sensitive to changes in carbon-dioxide-induced hydrogen ion concentration in the brain extracellular fluid.

Increased partial pressure of carbon dioxide within the brain extracellular fluid causes a corresponding increase in the concentration of hydrogen ions.

An elevation in the hydrogen ion concentration in the brain extracellular fluid directly stimulates the central chemoreceptors, which in turn increase ventilation by stimulating the respiratory center.

A decline in arterial partial pressure of carbon dioxide below normal is paralleled by a fall in partial pressure of carbon dioxide and hydrogen ions in the brain extracellular fluid, the result of which is a central-chemoreceptor-mediated decrease in ventilation. The powerful influence of the central chemoreceptors on the respiratory center is responsible for your inability to deliberately hold your breath for more than about a minute.

During apnea, a person subconsciously "forgets to breathe," whereas during dyspnea, a person consciously feels that ventilation is inadequate (page 356).

Apnea is the transient cessation of ventilation with the expectation that breathing will resume spontaneously.

The condition is called respiratory arrest.

In exaggerated cases of sleep apnea, the victim may be unable to recover from an apneic period, and death results.

This is the case of sudden infant death syndrome.

People with dyspnea have the subjective sensation that they are not getting enough air.

## KEY TERMS

Alveolar surface tension-p335
Anatomic dead space-p338
Aortic bodies-p355
Apneusis-p354
Apneustic center-p353
Apnea-p356
Asthma-p331
Bohr effect-p349
Bronchi-p323
Bronchioles-p323
Bronchoconstriction-p331
Bronchodilation-p331
Carbamino hemoglobin-p350
Carboxyhemoglobin-p349
Carotid bodies-p355
Central chemoreceptors-p356
Chronic bronchitis-p331
Chronic obstructive pulmonary disease-p331
Compliance-p333
Dorsal respiratory group-p353
Dyspnea-p357
Elastic recoil-p333
Emphysema-p332

Expiratory reserve volume-p337
Haldane effect-p350
Hypercapnia-p351
Hyperoxia-p351
Hyperpnea-p353
Hyperventilation-p351
Hypoxia-p351
Intrapleural fluid-p325
Newborn respiratory distress syndrome-p335
Oxygen-hemoglobin dissociation curve-p346
Oxyhemoglobin-p346
Partial pressure-p341
Pleurisy-p325
Pneumotaxic center-p353
Pneumothorax-p328
Pulmonary stretch receptors-p354
Pulmonary surfactant-p324
Residual volume-p337
Sudden infant death syndrome-p357
Type I alveolar cells-p324
Type II alveolar cells-p324
Ventral respiratory group-p354

# REVIEW EXERCISES

**True - False** (Answers on page A-23)

_____ 1. Pulmonary connective tissue contains large quantities of elastin fibers.

_____ 2. The work of breathing may be increased when pulmonary compliance is increased.

_____ 3. The maximum amount of air that the lungs can hold is about 4.2 liters in males.

_____ 4. Lungs will contain about 2,200 ml of air at the end of a normal quiet expiration

_____ 5. Alveolar ventilation is more important than pulmonary ventilation.

_____ 6. A gas always diffuses down its partial pressure gradient from the area of higher to the area of lower partial pressure.

_____ 7. The alveolar parial pressure of oxygen remains relatively constant at about 150 mm Hg throughout the respiratory cycle.

_____ 8. Surface area is reduced in emphysema.

_____ 9. The rate of gas transfer is directly proportional to the diffusion coefficient.

_____ 10. The amount of oxygen transferred to the cells and the amount of carbon dioxide carried away from the cells depend on the rate of cellular metabolism.

_____ 11 The net diffusion of oxygen occurs first between the blood and the tissues.

_____ 12. Respiratory control centers housed in the brain stem are responsible for generating the rhythmic pattern of breathing.

_____ 13. VRG is generally regarded as being responsible for the basic rhythm of ventilation.

_____ 14. The pneumotaxic center is dominant over the apneustic center.

_____ 15. Arterial partial pressure of carbon dioxide is the most important input regulating the magnitude of ventilation under resting conditions.

_____ 16. A fall in arterial partial pressure of carbon dioxide reflexly reduces the respiratory drive.

_____ 17. A decrease in hydrogen ion concentrations in the brain ECF directly stimulates the central chemoreceptors.

_____ 18. In sudden infant death syndrome (SIDS) babies two to five months are victims as a result of the immaturity of the respiratory control mechanisms.

_____ 19. Only 2.5 percent of the carbon dioxide in the blood is dissolved.

_____ 20. Each hemoglobin molecule can carry up to four molecules of oxygen.

_____ 21. In the systemic capillaries the blood partial pressure of carbon dioxide is increased.

_____ 22. At a normal partial pressure of oxygen of 100 mm Hg, hemoglobin is 98.5 percent saturated.

_____ 23. Local elevation in temperature enhances oxygen release from hemoglobin for use by the more active tissues.

_____ 24. Carbon dioxide binds with the globin portion of hemoglobin.

_____ 25. Histotoxic hypoxia arises when too little oxygenated blood is delivered to the tissues.

_____ 26. There are no muscles within the alveolar walls.

_____ 27. Air tends to move from a region of lower pressure to a region of higher pressure.

_____ 28. The intrapleural pressure is greater than the intra-alveolar pressure.

_____ 29. The stretched lungs have a tendency to pull inward.

_____ 30. During inspiration, the intrapleural pressure increases.

_____ 31. At resting expiration, the intra-alveolar pressure decreases.

_____ 32. Expiration is normally a passive process.

_____ 33. If the blood flow is greater than the airflow to a given alveolus, the oxygen level in the alveolus and surrounding tissues will fall below normal.

**Fill in the Blank** (Answers on page A-23)

34. _____ oxygen delivery to the tissues is normal, but the cells are unable to use the oxygen available to them.

35. _____ refers to excess carbon dioxide in the arterial blood that is caused by hypoventilation.

36. The relationship between partial pressure of oxygen and percent hemoglobin saturation is depicted by an S-shaped curve known as the _____.

37. The influence of carbon dioxide and acid on the release of oxygen is known as the _____.

38. The combination of carbon monoxide and hemoglobin is known as _____ _____.

39. The fact that removal of oxygen from hemoglobin increases the ability of hemoglobin to pick up $CO_2$ and $CO_2$-generated $H^+$ is known as the _____.

40. _____ is characterized by a low arterial blood partial pressure of oxygen accompanied by inadequate hemoglobin saturation.

41. _____ is characterized by inflammatory fluid accumulation within or around the alveoli.

42. The individual pressure exerted independently by a particular gas within a mixture of gases is known as its _____.

43. A difference in partial pressure between pulmonary blood and alveolar air is known as _____.

44. Across systemic capillaries. Oxygen partial pressure gradient from blood to tissue cell is _____ and carbon dioxide partial pressure gradient from tissue cell to blood is _____.

45. _____ is the volume of air in the lungs at the end of a normal passive expiration.

46. Two general categories of respiratory dysfunction yield abnormal results during spirometry; _____ and _____ lung disease.

47. Not all of the inspired air gets down to the site of gas exchange in the alveoli. Part of it remains in the conducting airways, where it is not available for gas exchange. The volume is considered to be _____.

48. _____ is the pressure exerted by the weight of the air in the atmosphere on objects on the earth's surface.

49. _____ refers to decreased airway resistance by promoting bronchiolar smooth-muscle relaxation.

50. _____ is characterized by collapse of the smaller airways and a breakdown of alveolar walls.

51. _____ refers to how much effort is required to stretch or distend the lungs.

52. _____ is the volume of air entering or leaving the lungs during a single breath.

53. The primary respiratory control center is the _____.

54. There are two other respiratory centers higher in the brain stem in the pons:
    (1) _____ and (2) _____.

55. The _____ consists mostly of inspiratory neurons.

56. _____ is when the breathing pattern consists of prolonged inspiratory gasps abruptly interrupted by very brief expirations.

57. Arterial partial pressure of oxygen is monitored by peripheral chemoreceptors known as the _____ and _____.

58. _____ is the transient cessation of ventilation with the expectation that breathing will resume spontaneously.

59. People who have _____ have the subjective sensation that they are not getting enough air.

60. _____ involves the sum of the processes that accomplish ongoing passive movement of oxygen from the atmosphere to the tissues to support cellular metabolism.

61. Two tubes lead from the pharynx--the _____ through which air is conducted to the lungs, and the _____ the tube through which food passes to the stomach.

62. _____ are two bands of elastic tissue that lie across the opening of the larynx, which can be stretched and positioned in different shapes by laryngeal muscles.

63. The alveolar walls consist of a single layer of flattened _____.

64. _____ secrete pulmonary surfactant.

**Matching:** Match the description of the lung volume or lung capacity to the name of lung volume or capacity (Answers on page A-24)

_____ 65. The vital capacity plus the residual volume

_____ 66. The minimum volume of air remaining in the lungs even after maximal expiration

_____ 67. The extra volume of air that can be maximally inspired over and above the typically resting tidal volume

_____ 68. The maximum volume of air that the lungs can hold

_____ 69. The tidal volume plus the inspiratory reserve volume

_____ 70. The volume of air in the lungs at the end of a normal passive expiration

_____ 71. The volume of air entering or leaving the lungs during a single breath

_____ 72. The extra volume of air that can be actively expired by maximal contraction of the expiratory muscles beyond that normally passively expired at the end of a typical resting tidal volume

_____ 73. The tidal volume plus the inspiratory reserve volume plus the expiratory reserve volume

_____ 74. The expiratory reserve volume plus the residual volume

_____ 75. The maximum volume of air that can be inspired at the end of a normal quiet expiration

_____ 76. The maximum volume of air that can be moved out during a single breath following a maximal inspiration

a. total lung capacity
b. tidal volume
c. vital capacity
d. functional residual capacity
e. inspiratory reserve volume
f. residual volume
g. inspiratory capacity
h. expiratory reserve volume

**Multiple Choice** (Answers on page A-24)

_____ 77. Which condition is characterized by the person "forgetting to breathe?"

    a. newborn respiratory distress syndrome
    b. atelectasis
    c. sudden infant death syndrome
    d. dyspnea
    e. hypercapnia

_____ 78. Which of the following conditions is characterized by air entering the pleural cavity?
- a. apnea
- b. pleurisy
- c. dyspnea
- d. asthma
- e. pneumothorax

_____ 79. Which condition refers to insufficient oxygen at the cellular level?
- a. hypoxia
- b. hypercapnia
- c. hyperpnea
- d. emphysema
- e. atelectasis

_____ 80. Which of the following conditions indicates that the partial pressure of carbon dioxide is above normal in arterial blood?
- a. hyperoxia
- b. hypercapnia
- c. hyperpnea
- d. atelectasis
- e. hypercarbonium

_____ 81. Which of the following is classified as a chronic obstructive pulmonary disease?
- a. pneumothorax
- b. pleurisy
- c. atelectasis
- d. hypoxia
- e. asthma

_____ 82. Which of the following refers to an inflammation of the pleural sac?
- a. pleuritis
- b. pneumothorax
- c. pleurisy
- d. emphysema
- e. hyperpnea

_____ 83. Which of the following is classified as a chronic obstructive pulmonary disease?
- a. pleuritis
- b. pleurisy
- c. sleep apnea
- d. pneumothorax
- e. emphysema

_____ 84. Which of the following conditions refers to a situation in alveolar surface tension due to insufficient pulmonary surfactant?
- a. sudden infant death syndrome
- b. newborn respiratory distress syndrome
- c. pleurisy
- d. pleuritis
- e. pulmonary insurfactantosis

_____ 85. Which of the following conditions is characterized by transient cessation of ventilation?
- a. apnea
- b. hypocapnia
- c. hyperoxia
- d. hyperdypsnia
- e. atelectasis

_____ 86. Which of the following conditions refers to increased ventilation to meet increased oxygen needs?
- a. hypopnea
- b. apnea
- c. dyspnea
- d. hyperpnea
- e. hypoxia

_____ 87. Which of the following conditions is characterized by inflammation of the lower respiratory airways?
- a. dyspnea
- b. chronic bronchitis
- c. asthma
- d. pleuritis
- e. pleurisy

_____ 88. Which of the following conditions is a subjective sensation of not getting enough air?
- a. dyspnea
- b. emphysema
- c. apnea
- d. hyperpnea
- e. pleurisy

**Points to Ponder**

1. The lung of a frog is quite different when compared to that of a human. The frog lung is a thin cylindrical tube much like a finger on a rubber glove. How does this very simple structure adequately serve the frog whereas humans require millions of alveoli to survive?

2. How would you explain the position of the trachea being ventral to the esophagus: the food must pass over the trachea to be swallowed.

3. What should you do when a small child declares that he will hold his breath until he dies if his demands are not met.

## Clinical Perspectives

1. How do you avoid carbon monoxide in a closed environment?
2. How would you relate drowning to this chapter?
3. How does taking a dive in a hyperbaric chamber enhance treatment against certain anaerobic bacteria?
4. You have just come upon an accident victim. After summoning help you see that the victim has a pneumothorax. How could you possibly help this person until medical help arrives?

## Experiment of the Day

1. Exhale normally, then exhale a little more and a little more until you can no longer exhale. From which of the lung volumes or capacities are you exhaling? Repeat using inhalation until you can no longer inhale. Into which lung volumes or capacities are you inhaling?

2. Do some exercise until you are ventilating rapidly. Place a bag over your mouth and nose. What are the results and why?

# URINARY SYSTEM 11

## CHAPTER OVERVIEW

The urinary system, more specifically, the kidneys are major contributors to the phenomenon of multicellular organisms. The existence of multicellular organisms depends upon the presence of the extracellular fluid. Since virtually all cells interface with the extracellular fluid, the homeostasis of this fluid is of utmost importance. For the most part, it is the role of the kidney to provide this homeostasis.

To fulfill this role the kidneys regulate most extracellular fluid ions; maintain pH, plasma volume, water balance, osmolarity; eliminate waste products of metabolism and many foreign compounds; and in a secondary capacity; produce erythropoietin and convert vitamin D into its active form.

While the functions of the kidneys can be listed quite simply in a short paragraph the physiology of the kidney is extremely complex. The kidneys are true guardians of homeostasis in the multicellular organism.

## CHAPTER OUTLINE

### Introduction (Text page 362)

The kidneys perform a variety of functions aimed at maintaining homeostasis (page 362).

The simplest forms of life live in an external environment of fixed composition, the sea.

The individual cells of complex multicellular organisms are able to function and survive only in a fluid environment of essentially constant composition similar to the sea.

This salty internal fluid environment is the extracellular fluid that bathes all the cells of the body and must be homeostatically maintained.

Terrestrial animals are able to live on dry land independent of the sea because of their kidneys, the organs that, in concert with the hormonal and neural inputs that control their function, are primarily responsible for maintaining the stability of extracellular fluid volume and electrolyte composition.

The kidneys are the primary route for elimination of potentially toxic metabolic wastes and foreign compounds from the body.

The kidneys make adjustments in urinary output of water, salt, and other electrolytes to compensate for their abnormal losses through heavy sweating, vomiting, diarrhea, or hemorrhage.

The following are specific functions performed by the kidneys: (1) maintain water balance in the body; (2) regulate the quality and concentration of most ECF ions; (3) maintain proper plasma volume; (4) assist in maintaining the proper acid-base balance; (5) maintain proper osmolarity; (6) excrete end products of bodily metabolism; (7) excrete many foreign compounds; (8) secrete erythropoietin; (9) secrete renin; and (10) convert vitamin D into its active form.

## The kidneys form urine; the remainder of the urinary system is ductwork that carries the urine to the outside (page 363).

The urinary system consists of the urine-forming organs--the kidneys--and the structures that carry the urine from the kidneys to the outside for elimination from the body.

The kidney acts on the plasma flowing through it to produce urine, conserving materials to be retained in the body and eliminating unwanted materials into the urine.

## The nephron is the functional unit of the kidney (page 365).

Each kidney is composed of about one million microscopic functional units known as nephrons.

A nephron is the smallest unit capable of urine formation.

The arrangement of nephrons within the kidneys give rise to two distinct regions; an outer granular-appearing renal cortex, and an inner region, the renal medulla.

Each nephron consists of a vascular component and a tubular component.

The dominant portion of the vascular component is the glomerulus.

On entering the kidney, the renal artery systematically subdivides to ultimately form many small vessels known as afferent arterioles, one of which supplies each nephron.

The afferent arteriole delivers blood to the glomerular capillaries, which rejoin to form another arteriole, the efferent arteriole.

No oxygen or nutrients are extracted from the blood for use by the kidney tissues nor are waste products picked up from the surrounding tissue.

The efferent arteriole quickly divides into a second set of capillaries, the peritubular capillaries, which supply the renal tissue with blood and are important in exchanges between the tubular system and blood during conversion of the filtered fluid into urine.

The peritubular capillaries rejoin to form venules that ultimately drain into the renal vein, by which the blood leaves the kidney.

The tubular component begins with Bowman's capsule.

From Bowman's capsule, the filtered fluid passes into the proximal tubule, which lies entirely in the cortex.

The next segment, the loop of Henle, dips into the renal medulla.

The descending limb of Henle's loop plunges from the cortex into the medulla; the ascending limb traverses back up into the cortex.

The ascending limb passes through the fork formed by the afferent and efferent arterioles.

Both the tubular and vascular cells at this point are specialized to form the juxtaglomerular apparatus.

Beyond the juxtaglomerular apparatus, the tubule once again becomes highly coiled to form the distal tubule.

The distal tubule empties into a collecting tubule.

Each collecting duct or tubule plunges down through the medulla to empty its fluid contents into the renal pelvis.

The three basic renal processes are glomerular filtration, tubular reabsorption, and tubular secretion (page 366).

Glomerular filtration is the first step in urine formation.

The filtrate flows through the tubules.

Selective movement of substances from inside the tubule into the blood is referred to as tubular reabsorption.

Tubular secretion refers to the selective transfer of substances from the peritubular capillary blood into the tubular lumen.

Urine excretion refers to the elimination of substances from the body in the urine and is the result of the first three processes.

# Glomerular Filtration (Text page 367)

The glomerular membrane is more than one hundred times more permeable than capillaries elsewhere (page 367).

Fluid filtered from the glomerulus into Bowman's capsule must pass through three layers that make up the glomerular membrane: (1) the wall of the glomerular capillaries, (2) an acellular gelatinous layer known as the basement membrane, and (3) the inner layer of Bowman's capsule.

The glomerular capillary wall is perforated by many large pores.

The final layer consists of podocytes, octopuslike cells that encircle the glomerular tuft.

The glomerular capillary blood pressure is the major force responsible for inducing glomerular filtration (page 369).

No active transport mechanisms or local energy expenditures are involved in moving fluid from the plasma across the glomerular membrane.

Passive physical forces are responsible for glomerular filtration.

Three physical forces are involved in glomerular filtration: (1) glomerular capillary blood pressure, (2) plasma colloid osmotic pressure, and (3) Bowman's capsule hydrostatic pressure.

The glomerular capillary blood pressure is higher than capillary blood pressure elsewhere.

Plasma-colloid osmotic pressure is caused by the unequal distribution of plasma proteins across the glomerular membrane.

The resultant tendency for water to move by osmosis down its own concentration gradient from Bowman's capsule into the glomerulus opposes glomerular filtration.

The fluid in Bowman's capsule exerts a hydrostatic pressure.

This pressure, which tends to push fluid out of Bowman's capsule, opposes glomerular filtration.

The total force favoring filtration is attributable to the glomerular capillary blood pressure.

The total of the two forces opposing filtration is less.
The net difference favoring filtration is referred to as the net filtration pressure.

## The most common factor resulting in a change in the GFR is an alteration in the glomerular capillary blood pressure by means of the sympathetic nervous system (page 370).

Plasma colloid osmotic pressure and Bowman's capsule hydrostatic pressure are not subject to regulation and under normal conditions do not vary substantially.
Glomerular capillary blood pressure can be controlled to adjust the GFR to suit the body's needs.
The magnitude of the glomerular capillary blood pressure depends on the rate of blood flow in each of the glomeruli and this flow is determined by the resistance offered by the afferent arterioles.
Changes in the GFR are accomplished by the sympathetic nervous system regulating the caliber of the afferent arterioles.
The parasympathetic nervous system does not exert any influence on the kidneys.
Sympathetic control of GFR is aimed at long-term regulation of arterial blood pressure.
A fall in arterial blood pressure is detected by the arterial carotid sinus and the aortic baroreceptors, which initiate neural reflex to increase blood pressure.

## The kidneys normally receive 20% to 25% of the cardiac output (page 371).

Most of the blood goes to the kidneys not to supply the renal tissue but to be adjusted and purified by the kidneys.
Only by continuously processing such a large proportion of the blood are the kidneys able to precisely regulate the volume and electrolyte composition of the internal environment and adequately eliminate large quantities of metabolic waste products that are constantly produced.

# Tubular Reabsorption (Text page 372)

## Tubular reabsorption is tremendous, highly selective, and variable (page 372).

Tubular reabsorption is a highly selective process.
The tubules have a high reabsorptive capacity for substances needed by the body and a poor or no reabsorptive capacity for substances of no value.
Only a small percentage of filtered plasma constituents that are useful to the body are present in the urine, most having been reabsorbed and returned to the blood.
Only excess amounts of essential materials such as electrolytes are excreted in the urine.
As water and other valuable constituents are reabsorbed, the waste products remaining in the tubular fluid become highly concentrated.

## Tubular reabsorption involves transepithelial transport (page 372).

To be reabsorbed, a substance must traverse five distinct barriers: (1) It must leave the tubular fluid by crossing the luminal membrane of the tubular cell. (2) It must pass through the cytosol from one side of the tubular cell to the other. (3) It must traverse the basolateral membrane of the tubular cell to enter the interstitial fluid. (4) It must diffuse through the interstitial fluid. (5) It must penetrate the capillary wall to enter the blood plasma.
There are two types of tubular reabsorption---passive reabsorption and active reabsorption.

An energy-dependent sodium-potassium ATPase transport mechanism in the basolateral membrane is essential for sodium reabsorption (page 373).

Sodium reabsorption in the proximal tubule plays a pivotal role in the reabsorption of glucose, amino acids, water chloride ions and urea.

Sodium reabsorption in the loop of Henle, along with chloride ion reabsorption, plays a critical role in the kidney's ability to produce urine of varying concentrations and volumes, depending on the body's need to conserve or eliminate water.

Sodium reabsorption in the distal portions of the nephron is variable and subject to hormonal control, being important in the regulation of ECF volume.

Sodium reabsorption is also linked in part with potassium secretion.

Aldosterone stimulates sodium reabsorption in the distal and collecting tubules; atrial natriuretic peptide inhibits it (page 374).

The most important and best known hormonal system involves in the regulation of sodium is the renin-angiotension-aldosterone system, which stimulates sodium reabsorbtion in the distal and collecting tubules.

The juxtaglomerulus apparatus secrete a hormone, renin into the blood.

Renin acts as an enzyme to activate angiotensinogen into angiotensin I.

On passing through the lungs, angiotensin I is converted to angiotensin II.

Angiotensin II is the primary stimulus for the secretion of the hormone aldosterone.

Aldosterone increases sodium reabsorption by the distal and collecting tubules.

The renin-angiotensin-aldosterone system thus promotes salt retention and a resultant water retention and elevation of arterial blood pressure.

Atrial natriuretic peptide is released from the cardiac atria when the ECF volume is expanded.

The primary action of atrial natriuretic peptide is to inhibit sodium reabsorption in the distal parts of the nephron.

Glucose and amino acids are reabsorbed by sodium-dependent secondary active transport (page 376).

Glucose and amino acids are transferred by means of secondary active transport; a specialized cotransport carrier simultaneously transfers both sodium and the specific organic molecule from the lumen into the cell.

Glucose and amino acids get a "free ride" at the expense of energy already used in the reabsorption of sodium.

With the exception of sodium, actively reabsorbed substances exhibit a tubular maximum (page 376).

The quantity of any substance filtered per minute is known as the filtered load.

At a constant GFR, the glucose filtered load is directly proportional to the plasma glucose concentration.

The plasma glucose concentration can become extremely high in diabetes mellitus.

Consequently, although glucose does not normally appear in the urine, it is found in the urine of persons with diabetes when the plasma glucose concentration exceeds the renal threshold.

The kidneys do directly contribute to the regulation of many electrolytes such as calcium and phosphate.

Our diets are generally rich in phosphates, but because the tubules can reabsorb up to the normal plasma concentration's worth of phosphate and no more, the excess ingested phosphate is quickly spilled into the urine.

Unlike the reabsorption of organic nutrients, the reabsorption of phosphate and calcium is also subject to hormonal control.

Parathyroid hormone can alter the renal thresholds for phosphate and calcium, thus adjusting the quantity of these electrolytes conserved depending on the body's momentary needs.

## Active sodium reabsorption is responsible for the passive reabsorption of chloride ions, water, and urea (page 378).

The negatively charged chloride ions are passively reabsorbed down the electrical gradient created by the active reabsorption of positively charged sodium ions.

Water is passively reabsorbed by osmosis throughout the length of the tubule.

The passive reabsorption of urea is also indirectly linked to active sodium reabsorption.

## In general, unwanted waste products are not reabsorbed (page 378).

Urea molecules, being the smallest of waste products, are the only wastes able to be passively reabsorbed as a result of the concentrating effect of the kidneys.

The waste products, failing to be reabsorbed, generally remain in the tubules and are excreted in the urine in a highly concentrated form.

# Tubular Secretion (Text page 378)

## The most important secretory processes are those for hydrogen, potassium, and organic ions (page 378).

The most important substances secreted by the tubules are hydrogen ions, potassium ions, organic anions and cations, many of which are compounds foreign to the body.

Renal hydrogen ion secretion is extremely important in the regulation of acid-base balance in the body.

Hydrogen ions can be added to the filtered fluid by being secreted by the proximal, distal, and collecting tubules.

Potassium is actively reabsorbed in the proximal tubule and actively secreted in the distal and collecting tubules.

Potassium ion secretion in the distal and collecting tubules is coupled to sodium ion reabsorption by means of the energy-dependent basolateral sodium-potassium pump.

Increased aldosterone secretion always promotes simultaneous sodium reabsorption and potassium secretion.

The proximal tubule contains two distinct types of secretory carriers, one for the secretion of organic anions and a separate system for the secretion of organic cations.

# Urine Excretion and Plasma Clearance (Text page 380)

## On the average, one milliliter of urine is excreted per minute (page 380).

Typically, of the 125 ml of plasma filtered per minute, 124 ml/min are reabsorbed, so the final quantity of urine formed averages 1 ml/min.

Thus, 1.5 liter of urine per day are excreted, out of the 180 liters per day filtered.

A relatively small change in the quantity of filtrate reabsorbed can bring about a large change in the volume of urine formed.

Plasma clearance refers to the volume of plasma cleared of a particular substance per minute (page 381).

> If a substance is filtered but not reabsorbed or secreted, its plasma clearance rate equals the GFR.
>
> If a substance is filtered and reabsorbed but not secreted, its plasma clearance rate is always less than the GFR.
>
> If a substance is filtered and secreted but not reabsorbed, its plasma clearance rate is always greater than the GFR.

The ability to excrete urine of varying concentrations depends on the medullary countercurrent system and vasopressin (page 382).

> The ECF osmolarity depends on the relative amount of water compared to solute.
>
> At normal fluid balance and solute concentration, the body fluids are said to be isotonic.
>
> If there is too much water relative to the solute load, the body fluids are hypotonic.
>
> If a water deficit exists relative to the solute load, the body fluids are too concentrated or are hypertonic.
>
> A large vertical osmotic gradient is uniquely maintained in the interstitial fluid of the medulla of each kidney.
>
> The concentration of the interstitial fluid progressively increases from the cortical boundary down through the depth of the renal medulla until it reaches a maximum of 1,200 mosm/liter in humans at the junction with the renal pelvis.
>
> The presence of this gradient enables the kidneys to produce urine that ranges in concentration from 100 to 1,200 mosm/ liter.
>
> When the body is in ideal fluid balance, isotonic urine is formed.
>
> When the body is overhydrated, the kidneys are able to produce a large volume of dilute urine thus eliminating the excess water in the urine.
>
> Conversely, the kidneys are able to put out a small volume of concentrated urine when the body is dehydrated.
>
> The long loops of Henle establish the vertical osmotic gradient.
>
> The collecting tubules of all nephrons use the gradient, in conjunction with the hormone vasopressin, to produce urine of varying concentrations.
>
> Collectively, this entire functional organization is known as the medullary countercurrent system.
>
> The long loops of Henle establish the medullary vertical osmotic gradient by means of countercurrent multiplication.
>
> The descending limb is highly permeable to water but does not actively extrude sodium.
>
> The ascending limb actively transports NaCl out of the tubular lumen into the surrounding interstitial fluid and is always impermeable to water, so salt leaves the tubular fluid without water osmotically following along.
>
> The vertical osmotic gradient in the medullary interstitial fluid is used by the collecting ducts to concentrate the tubular fluid so that a urine more concentrated than normal body fluids can be excreted.
>
> The fact that the fluid is hypotonic as it enters the distal portions of the tubule enables the kidneys to excrete a urine more dilute than normal body fluids.
>
> In order for water reabsorption to occur across a segment of the tubule, two criteria must be met: (1) an osmotic gradient must exist across the tubule, and (2) the tubular segment must be permeable to water.

The distal and collecting tubules are impermeable to water except in the presence of vasopressin, also known as antidiuretic hormone.

Vasopressin is produced in the hypothalamus and is stored in the posterior pituitary gland.

The hypothalamus controls the release of vasopressin from the posterior pituitary into the blood.

In negative-feedback fashion, vasopressin secretion is stimulated by a water deficit when water must be conserved for the body and inhibited by a water excess when surplus water must be eliminated in the urine.

## Renal failure has wide-ranging consequences (page 389).

When the functions of both kidneys are disrupted to the point that they are unable to perform their regulatory and excretory functions sufficiently to maintain homeostasis, renal failure is said to exist.

Renal failure can manifest itself either as acute renal failure or as chronic renal failure.

## Urine is temporarily stored in the bladder, from which it is emptied by the process of micturition (page 389).

Peristaltic contractions of the smooth muscles within the ureteral wall propel the urine forward from the kidneys to the bladder.

As the bladder fills, the ureteral ends within its walls are compressed closed.

Urine can still enter because ureteral contractions generate sufficient pressure to overcome the resistance.

As is characteristic of smooth muscle, bladder muscle is able to stretch tremendously without a build-up in bladder-wall tension.

The bladder smooth muscle is richly supplied by parasympathetic fibers, stimulation of which causes bladder contraction.

The exit from the bladder is guarded by two sphincters; the internal urethral sphincter and the external urethral sphincter.

The internal urethral sphincter is under involuntary control.

The external urethral sphincter and pelvic diaphragm are under voluntary control.

Micturition, the process of bladder emptying, is governed by two mechanisms:  the micturition reflex and voluntary control.

## KEY TERMS

| | | |
|---|---|---|
| Active reabsorption-p373 | Hypertonic-p382 | Peritubular capillaries-p366 |
| Afferent arterioles-p365 | Hypotonic-p382 | Plasma clearance-p381 |
| Bowman's capsule-p366 | Isotonic-p382 | Proximal tubule-p366 |
| Collecting tubule-p366 | Juxtaglomerular apparatus-p366 | Renal failure-p388 |
| Distal tubule-p366 | Loop of Henle-p366 | Renal threshold-p377 |
| Diuretics-p375 | Micturition-p391 | Tubular reabsorption-p366 |
| Efferent arterioles-p365 | Micturition reflex-p391 | Tubular secretion-p367 |
| Glomerular filtration-p366 | Net filtration pressure-p369 | Urea-p378 |
| Glomerulus-p365 | Passive reabsorption-p373 | Vasopressin-p385 |

## REVIEW EXERCISE

**True - False** (Answers on page A-25)
_____ 1. The afferent arteriole delivers blood to the glomerular capillaries.
_____ 2. The efferent arteriole carries venous blood to the renal vein to exit the kidney.
_____ 3. The peritubular capillaries supply the renal tissue with oxygen.
_____ 4. The glomerulus in a juxtamedullary nephron is located in the medulla of the kidney.
_____ 5. Podocytes are found in the membranes of the distal tubule.
_____ 6. The GFR is dependent upon the net filtration pressure only.
_____ 7. Aldosterone stimulates sodium reabsorption in the collecting tubules.
_____ 8. Atrial natriuretic peptide is released from the juxtaglomerular apparatus.
_____ 9. The filtered load is known as the level at which the tubular maximum has been reached and the particular substance begins to appear in the urine.
_____ 10. Negatively charged chloride ions are passively reabsorbed in the proximal tubule.
_____ 11. Urea is passively reabsorbed in the distal and collecting tubule.
_____ 12. Organic cations are secreted by the proximal tubules.
_____ 13. Sodium is passively reabsorbed in the proximal tubule.
_____ 14. The most important substances secreted by the distal and collecting tubules are sodium and potassium.
_____ 15. Potassium is actively reabsorbed in the proximal tubule.
_____ 16. Potassium is actively secreted in the distal and collecting tubules.
_____ 17. Urine excretion is about 10.5 liters per day.
_____ 18. If a substance is filtered and reabsorbed but not secreted, its plasma clearance is always greater than the GFR.
_____ 19. If a substance is filtered and secreted but not reabsorbed, its plasma clearance is always less than the GFR.
_____ 20. If a substance is filtered but not secreted or reabsorbed, its plasma clearance rate equals the GFR.
_____ 21. Vasopressin is produced in the posterior pituitary and is also known as antidiuretic hormone.

**Fill in the Blank** (Answers on page A-25)
22. _____ is a waste product resulting from the breakdown of proteins.
23. The passive reabsorption of urea is also directly linked to active _____ reabsorption.
24. The parathyroid hormone can alter renal thresholds for _____.
25. Atrial natriuretic peptide is released from the cardiac atria when the ECF volume _____.

26. Fluid retention and edema that accompany congestive heart failure are brought about by _____ activity.
27. Renin acts as an enzyme to activate angiotensinogen into _____.
28. _____ is the primary stimulus for the secretion of aldosterone.
29. _____ increases sodium reabsorption by the distal and collecting tubules.
30. Tubular reabsorption involves _____ transport.
31. Each tuft of glomerular capillaries is held together by _____.
32. The rate of glomerular filtration is dependent on the _____ as well as on the _____.
33. From Bowman's capsule, the filtered fluid passes into the _____ tubule.
34. Solute excretion is always accompanied by comparable water excretion because of _____ considerations.
35. The early portion of the collecting tubule is _____ to urea.
36. _____ contractions of the smooth muscle within the ureteral wall propel urine forward from the kidneys to the bladder.
37. The exit from the bladder is guarded by two sphincters: _____ and _____.
38. Body fluids are said to be isotonic at an osmolarity of _____ mosm/l.
39. Potassium-ion secretion in the distal and collecting tubules is coupled to sodium reabsorption by means of the energy-dependent basolateral _____.
40. Increased aldosterone secretion always promotes simultaneous sodium reabsorption and _____.

**Matching:** Using the code at the right, match the reabsorption or secretion function to all the tubules that perform the function (Answers on page A-26)

_____ 41. Organic-ion secretion; not subject to control
_____ 42. Variable water reabsorption, controlled by vasopressin
_____ 43. Variable hydrogen ion secretion, depending     a. proximal tubule
       on acid-base status of the body            b. distal tubule
_____ 44. All filtered glucose and amino acids reabsorbed by   c. collecting tubule
       secondary active transport, not subject to control
_____ 45. Variable sodium reabsorption, controlled by aldosterone; chloride follows passively
_____ 46. All filtered potassium reabsorbed; not subject to control
_____ 47. 50% of filtered urea passively reabsorbed; not subject to control
_____ 48. Variable amounts of phosphate ions and other electrolytes reabsorbed; subject to control
_____ 49. Variable potassium secretion
_____ 50. 67% of filtered sodium actively reabsorbed; not subject to control; chloride follows passively
_____ 51. 65% of filtered water osmotically reabsorbed; not subject to control

**Multiple Choice** (Answers on page A-26)

_____ 52. Which of the following substances is secreted by the kidneys and stimulates red blood cell production?
   a. aldosterone
   b. vasopressin
   c. para-aminohippuric acid
   d. renin
   e. erythropoietin

_____ 53. Which substance is secreted by the hypothalamus, stored in the posterior pituitary, and stimulates the reabsorption of water in the collecting tubules?
   a. aldosterone
   b. angiotensin I
   c. vasopressin
   d. para-aminohippuric acid
   e. inulin

_____ 54. Which of the following substances is used to clinically ascertain the GFR?
   a. aldosterone
   b. inulin
   c. renin
   d. atrial natriuretic peptide
   e. vasopressin

_____ 55. Which of the following substances is the primary stimulus for the secretion of the hormone aldosterone?
   a. angiotensin I
   b. renin
   c. inulin
   d. angiotensin II
   e. atrial natriuretic peptide

_____ 56. Which substance inhibits sodium reabsorption in the distal parts of the nephron?
   a. atrial natriuretic peptide
   b. para-aminohippuric acid
   c. vasopressin
   d. aldosterone
   e. renin

_____ 57. Which of the following substances is a hormone of the adrenal gland?
   a. renin
   b. aldosterone
   c. inulin
   d. vasopressin
   e. erythropoietin

_____ 58. Which substance functions as an enzyme to activate angiosinogen into angiotensin I ?
   a. aldosterone
   b. vasopressin
   c. inulin
   d. renin
   e. para-hippuric acid

_____ 59. Which of the following substances can be used to measure renal plasma flow?
   a. angiotensin I
   b. para-aminohippuric acid
   c. angiotensin II
   d. aldosterone
   e. vasopressin

_____ 60. Which of these substances must pass through the lungs before they can become a stimulus for hormone release?
   a. angiotensin I
   b. angiotensin II
   c. angiotensinogen
   d. inulin
   e. aldosterone

_____ 61. Which of the following substances increases sodium reabsorption by the distal tubule?
   a. renin
   b. angiotensin I
   c. para-aminohippuric acid
   d. atrial natriuretic peptide
   e. aldosterone

**Points to Ponder**
1. What is the advantage of sweat containing salt and other ions?  Would water work just as well?
2. What makes the urine yellow?
3. Mammals excrete urea, uric acid, and creatinine as liquids while other vertebrates excrete these waste products as solids.  Is there an advantage to this?
4. What would happen if we cut one ureter and insert the proximal end into a vein so that the urine flows directly back into the blood?  The other kidney functions normally.

**Clinical Perspective**
1. A favorite fraternity pledge prank is to put methyl blue in the coffee of your big brother.  Explain why the urine turns blue.

**Experiment of the Day**

1. Eat some asparagus. Note the time of ingestion. Drink plenty of fluids. During the next twenty-four hours check your urine for any noticeable changes. Note the time that the changes are first observed. Explain the changes should you become aware of any. If you observe no changes, ask a classmate for their changes.

# FLUID AND ACID-BASE
# BALANCE

## CHAPTER OVERVIEW

The theme continues.  As changes occur, homeostasis is maintained by body systems
responding to the changes.  This chapter deals with two very important areas of change:
water and pH.

Water is located primarily in two major compartments; extracellular fluid (plasma and
interstitial fluid) and intracellular fluid.  All of these have different amounts of water
and ions.  To regulate and maintain these fluid balances, adjustments are made in the
extracellular volume and the extracellular osmolarity.

Acid-base balance is very critical in the survival of cells.  The somewhat narrow pH
range is maintained by four buffer systems:  carbonic acid/carbonate buffer system,
protein buffer system, hemoglobin buffer system, and the phosphate buffer system.

The buffers are the first line of defense against pH changes.  The second line is the lungs.
By removing carbon dioxide less carbonic acid is added to body fluids.  The last line of
defense in the prevention of change in pH is the kidney.  The big role of the kidney is to
excrete the necessary amounts of hydrogen ions, bicarbonate ions, and ammonium ions.
Maintain a relatively constant pH is truly remarkable considering the great variability
of substances brought into and produced by the body.

## CHAPTER OUTLINE

### Fluid Balance (Text page 397)

Input must equal output if balance is to be maintained (page 397).

If the quantity of a substance is to remain stable within the body its input by means of
ingestion, or metabolic production must be balanced by an equal output by means of
excretion or metabolic consumption.

Known as the balance concept, this relationship is extremely important in the maintenance of homeostasis.

If the body as a whole has a surplus or a deficit of a particular stored substance, the storage site can be expanded or partially depleted to maintain the extracellular fluid concentration of the substance within homeostatically prescribed limits.

When total body input of a substance equals total body output, a stable balance exists.

When the gains via input exceed its losses via output, a positive balance exists.

When the losses of a substance exceed its gains, a negative balance exists.

## Body water is distributed between the intracellular and extracellular fluid compartments (page 398).

The intracellular fluid compartment comprises about two-thirds of the total body water.

The remaining one-third of the body water, found in the extracellular fluid, is further subdivided into plasma and interstitial fluid.

Two other minor categories are included in the extracellular fluid compartment: lymph and transcellular fluid.

Transcellular fluid consists of a number of small specialized fluid volumes.

Although these fluids are extremely important functionally, they represent an insignificant fraction of the total body water.

## The plasma and interstitial fluid are separated by the blood vessel walls, whereas ECF and ICF are separated by cellular plasma membranes (page 398).

Plasma and interstitial fluid are nearly identical in composition, except that interstitial fluid lacks plasma proteins.

Any change in one of these ECF compartments is quickly reflected in the other compartment because they are constantly mixing.

The composition of the ECF differs considerably from that of the ICF.

Among the major differences are: (1) the presence of cellular proteins in the ICF that are unable to permeate the enveloping membranes to leave the cells, and (2) the unequal distribution of sodium and potassium and their attendant anions as a result of the action of the membrane-bound sodium-potassium ATPase pump that is present in all cells.

Sodium is the primary ECF cation and potassium is primarily found in the ICF.

In the extracellular fluid, sodium is accompanied primarily by the anion chloride and to a lesser extent by bicarbonate.

The major intracellular anions are phosphate and the negatively charged proteins trapped within the cell.

## Fluid balance is maintained by regulating ECF volume and ECF osmolarity (page 399).

Any control mechanism that operates on the plasma in effect regulates the entire extracellular fluid.

The ICF, in turn, is influenced by changes in the ECF to the extent permitted by the permeability of the membrane barriers surrounding the cells.

The factors regulated to maintain fluid balance in the body are ECF volume and ECF osmolarity.

Extracellular fluid volume must be closely regulated to help maintain blood pressure.

Extracellular fluid osmolarity must be closely regulated to prevent swelling or shrinking of the cells.

## Control of ECF volume is important in the long-term regulation of blood pressure (page 400).

Two compensatory measures come into play to transiently adjust the blood pressure until the ECF volume can be restored to normal: (1) baroreceptor reflex mechanisms alter both cardiac output and total peripheral resistance through autonomic nervous system effects on the heart and blood vessels, and (2) fluid shifts occur temporarily and automatically between the plasma and interstitial fluid.

A reduction in plasma volume is partially compensated for by a shift of fluid out of the interstitial compartment into the blood vessels, thereby expanding the circulating plasma volume at the expense of the interstitial compartment.

Conversely, when the plasma volume is too large, much of the excess fluid is shifted into the interstitial compartment.

Responsibility for long-term regulation of blood pressure rests with the kidneys and the thirst mechanism, which control urinary output and fluid intake.

## Control of salt balance is primarily important in regulating ECF volume (page 400).

The total mass of sodium salts in the ECF determines the ECF's volume, and appropriately, regulation of the ECF volume depends primarily on controlling salt balance.

To maintain salt balance at a set level, salt input must equal salt output.

Since we typically consume salt in excess of our needs, it is obvious that salt intake in humans is not well controlled.

The three avenues for salt output are obligatory loss of salt in sweat and feces and controlled excretion of salt in the urine.

By regulating the rate of urinary salt excretion, the kidneys keep the total sodium mass in the ECF constant despite any notable changes in dietary intake of salt or unusual losses through sweating, diarrhea, or other means.

The amount of sodium excreted in the urine represents the amount of sodium that is filtered but is not subsequently reabsorbed.

The kidneys accordingly adjust the amount of salt excreted by controlling two processes: (1) the glomerular filtration rate, and (2) the tubular reabsorption of sodium.

The afferent arterioles that supply the renal glomeruli are constricted as part of the generalized vasoconstriction aimed at elevating a reduced blood pressure.

As a result of reduced blood flow into the glomeruli, the glomerular filtration rate decreases and accordingly, the amount of sodium and accompanying fluid that are filtered decreases.

Excretion of salt and fluid are diminished.

An elevation in ECF volume and arterial blood pressure is reflexly countered by a baroreceptor reflex response that leads to an increase in glomerular filtration rate, which in turn results in enhanced salt and fluid excretion.

Baroreceptors that monitor fluctuations in blood pressure are responsible for bringing about adjustments in the amount of sodium filtered and eventually excreted.

The main factor controlling the extent of sodium reabsorption in the distal and collecting tubule is the powerful renin-angiotension-aldosterone system.

A fall in arterial blood pressure brings about a twofold effect in the renal handling of sodium: (1) a reflex reduction in the glomerular filtration rate to decrease the amount of sodium filtered, and (2) a hormonally adjusted increase in the amount of sodium reabsorbed.

A rise in arterial blood pressure brings about (1) increases in the amount of sodium filtered, and (2) a reduction in renin-angiotension-aldosterone activity, which decreases salt and water reabsorption.

## Control of ECF osmolarity prevents changes in ICF volume (page 401).

Normally, the osmolarities of the ECF and ICF are the same.

Hypertonicity of the ECF, or the excessive concentration of ECF solutes, is usually associated with dehydration, or a negative free water balance.

Dehydration with accompanying hypertonicity can be brought about in three major ways: (1) insufficient water intake; (2) excessive water loss; and (3) diabetes insipidus.

Whenever the ECF compartment becomes hypertonic, water moves out of the cells by osmosis into more concentrated ECF until the ICF osmolarity equilibrates with the ECF. The cells shrink as water leaves them.

Shrinking of brain neurons causes disturbances in brain function.

Circulatory problems may range from slight reduction in blood pressure to circulatory shock and death.

Hypotonicity of the ECF is usually associated with overhydration; that is excess of free water is present.

Hypotonicity can arise in three ways; (1) Patients with renal failure who are unable to excrete a dilute urine become hypotonic when they consume relatively more water than solutes. (2) Hypotonicity can occur transiently in healthy people if water is rapidly ingested to such an excess that the kidneys are unable to respond quickly enough to eliminate the extra water. (3) Hypotonicity can occur when excess water without solute is retained in the body as a result of inappropriate secretion of vasopressin.

The resultant difference in osmotic activity between ECF and ICF induces water to move by osmosis from the more dilute ECF into the cells, with the cells swelling.

Pronounced swelling in brain cells also leads to brain dysfunction.

## Control of water balance by means of vasopressin and thirst is of primary importance in regulating ECF osmolarity (page 405).

Of the many sources of water input and output only two can be regulated to maintain water balance: (1) Control of water input by thirst and (2) control of water output in the urine by vasopressin.

A thirst center is located in the hypothalamus in close proximity to the vasopressin-secreting cells.

Thirst increases water input, whereas vasopressin, by reducing urine production, decreases water output.

Vasopressin and thirst are both stimulated by a free water deficit and suppressed by a free water excess.

The predominant excitatory input for both vasopressin secretion and thirst come from the hypothalamus osmoreceptors located near the vasopressin-secreting cells and thirst center.

Left atrial volume receptors monitor the blood pressure, which is a reflection of the ECF volume.

In response to a major reduction in ECF volume and arterial pressure, as during hemorrhage, the left atrial volume receptors reflexly stimulate both vasopressin secretion and thirst.

The outpouring of vasopressin and increased thirst lead to decreased urine output and increased fluid intake.

Vasopressin exerts a potent vasoconstrictor effect on arterioles in addition to having an effect on the kidney tubules.

Vasopressin and thirst are both inhibited when the ECF/plasma volume and arterial blood pressure are elevated.

Aldosterone controlled sodium reabsorption is the most important factor in regulating ECF volume, with the vasopressin and thirst mechanisms playing only a supportive role.

# Acid-Base Balance (Text page 407)

## Acids liberate free hydrogen ions whereas bases accept them (page 407).

Acids are a special group of hydrogen-containing substances that dissociate when in solution to liberate free hydrogen ions.

A base is a substance that can combine with a free hydrogen ion thus removing it from solution.

## The pH designation is used to express hydrogen ion concentration (page 408).

The pH equals the logarithm to the base 10 of the reciprocal of the hydrogen ion concentration.

The greater the hydrogen ion concentration the lower the pH.

Solutions having a pH less than 7.0 are considered to be acidic.

Solutions having a pH greater than 7.0 are considered to be basic or alkaline.

## Fluctuations in hydrogen-ion concentration have profound effects on body chemistry (page 408).

Changes in excitability of nerve and muscle cells are among the major clinical manifestations of pH abnormalities.

Hydrogen-ion concentration exerts a marked influence on enzyme activity.

Changes in hydrogen-ion concentration influence potassium levels in the body.

## Hydrogen ions are continually being added to the body fluids as a result of metabolic activities (page 409).

Normally hydrogen is continually being added to the body fluids from the three following sources: (1) carbonic acid formation, (2) inorganic acids produced during the breakdown of nutrients, and (3) organic acids resulting from intermediary metabolism.

Three lines of defense against changes in hydrogen ion concentration operate to maintain the hydrogen ion concentration of body fluids at a nearly constant level despite unregulated input: (1) the chemical buffer systems, (2) the respiratory mechanism of pH control, and (3) the renal mechanism of pH control.

## Chemical buffer systems act as the first line of defense against changes in hydrogen ion concentration (page 410).

A chemical buffer system is a mixture in a solution of two chemical compounds that minimize pH changes when either an acid or a base is added or removed from the solution.

There are four buffer systems in the body: (1) the carbonic acid/bicarbonate buffer system, (2) the protein buffer system, (3) the hemoglobin buffer system, and (4) the phosphate buffer system.

The carbonic acid/bicarbonate buffer pair is the most important buffer system in the ECF for buffering pH changes brought about by causes other than fluctuations in carbon dioxide-generated carbonic acid.

The protein buffer system is primarily important intracellularly.

The hemoglobin buffer system buffers hydrogen ions generated from carbonic acid.

The phosphate buffer system is an important urinary buffer.

The phosphate buffer system is composed of an acid phosphate salt and a basic phosphate salt.

This system contributes significantly to intracellular buffering.

The phosphate system serves as an excellent urinary buffer.

All chemical buffer systems act immediately, within fractions of a second, to minimize changes in pH.

The respiratory system, as the second line of defense, regulates hydrogen-ion concentration by controlling the rate of carbon dioxide removal from the plasma through adjustments in pulmonary ventilation (page 412).

When arterial hydrogen ion concentration increases, the respiratory center in the brain stem is reflexly stimulated to increase pulmonary ventilation.

Since carbon dioxide forms acid, the removal of carbon dioxide removes acid from the body.

The kidneys, as the third line of defense, contribute powerfully to control of acid-base balance by controlling both hydrogen ion and bicarbonate concentrations in the blood (page 412).

The kidneys control the pH of the body fluids by adjusting three interrelated factors: (1) hydrogen excretion, (2) bicarbonate excretion, and (3) ammonia secretion.

The magnitude of hydrogen secretion depends on a direct effect of the plasma acid-base status on the kidney's tubular cells.

When the hydrogen ion concentration of the plasma passing through the peritubular capillaries is elevated above normal, the tubular cells respond by secreting greater-than-usual amounts of hydrogen from the plasma into the tubular fluid to be excreted in the urine.

When the plasma hydrogen ion concentration is lower than normal, the kidneys conserve hydrogen by reducing their secretion and subsequent excretion in the urine.

There are two important urinary buffers: (1) filtered phosphate buffers and (2) secreted ammonia.

Normally, secreted hydrogen is first buffered by the phosphate buffer system.

When acidosis exists, the tubular cells secrete ammonia into the tubular fluid once the normal urinary phosphate buffers are saturated.

Ammonia enables the kidneys to continue secreting additional hydrogen ions because ammonia combines with free hydrogen in the tubular fluid to form ammonium ions.

The ammonium ions remain in the tubular fluid and are lost in the urine, each one taking a hydrogen ion with it.

Acid-base imbalances can arise from either respiratory dysfunction or metabolic disturbances (page 414).

Respiratory acidosis is the result of abnormal carbon dioxide retention arising from hypoventilation.

The primary defect in respiratory alkalosis is excessive loss of carbon dioxide from the body as a result of hyperventilation.

The following are the most common causes of metabolic acidosis: (1) severe diarrhea, (2) diabetes mellitus, (3) strenuous exercise, and (4) uremic acidosis.

Metabolic alkalosis arises most commonly from the following: (1) vomiting and (2) ingestion of alkaline drugs.

## KEY TERMS

Acidosis-p408

Acids-p408

Alkalosis-p408

Ammonia-p413

Ammonium ions-p413

Balance concept-p397

Bases-p408

Chemical buffer system-p410

Dehydration-p402

Diabetes insipidus-p402

Hypothalamic osmoreceptors-p521

Long atrial volume receptors-p406

Negative balance-p397

Osmolarity-p401

Overhydration-p405

Positive balance-p397

Stable balance-p397

Thirst center-p406

Transcellular fluid-p398

Water intoxication-p405

## REVIEW EXERCISES

**True - False** (Answers on page A-27)

_____ 1. When the gains via input exceed its losses via output, a negative balance exists.

_____ 2. The ICF compartment comprises about one-half of the total body water.

_____ 3. Plasma is a transcellular fluid.

_____ 4. The major intracellular anions are phosphate ions.

_____ 5. Diabetes insipidus is a disease characterized by a deficiency in vasopressin.

_____ 6. Vasopressin decreases water output by reducing urine production.

_____ 7. Vasopressin gets its name because it exerts a potent vasoconstrictor effect on arterioles.

_____ 8. Bases are a special group of hydrogen containing substances that dissociate when in solution to liberate free hydrogen.

_____ 9. A base is a substance that can combine with free hydrogen and thus remove it from solution.

_____ 10. The greater the hydrogen ion concentration the lower the pH.

_____ 11. The first line of defense against change in pH is the respiratory control of pH.

_____ 12. Tubular cells secrete ammonium ions that can combine with free hydrogen to form ammonia.

_____ 13. Metabolic acidosis can be caused by diabetes insipidus.

_____ 14. Metabolic alkalosis can be caused by vomiting.

**Fill in the Blank** (Answers on page A-27)

15. When total body input of a particular substance equals its total body output, a _____ exists.

16. _____ and _____ are nearly identical in composition, except that _____ lacks plasma proteins.

17. The three avenues for salt output are _____ _____, _____, and controlled excretion of salt in the _____.

18. The main factor controlling the extent of sodium reabsorption in the distal tubule and collecting tubule is the powerful _____ system.

19. Dehydration with accompanying hypertonicity can be brought about in three major ways: (1) _____, (2) _____ and (3) _____.

20. The predominant excitatory input for both vasopressin secretion and thirst comes from _____.

21. The _____ equals the logarithm to the base 10 of the reciprocal of the hydrogen ion concentration.

22. The _____ serves as an excellent urinary buffer.

23. Respiratory acidosis is the result of abnormal carbon dioxide retention arising from _____.

24. Uremic acidosis causes _____.

25. Vomiting causes _____.

26. Bicarbonate cannot buffer urinary hydrogen as it does in the ECF because bicarbonate is not excreted in the urine simultaneously with _____.

27. The kidneys control the pH of the body fluids by adjusting three interrelated factors: (1) _____, (2) _____ and (3) _____

**Matching:** Match the cause on the left to the acid-base condition on the right
        (Answers on page A-27)

_____ 28. diabetes mellitus                     a. metabolic alkalosis
_____ 29. hyperventilation                      b. respiratory acidosis
_____ 30. vomiting                              c. metabolic acidosis
_____ 31. ingesting of alkaline drugs           d. respiratory alkalosis
_____ 32. strenuous exercise
_____ 33. uremic acidosis
_____ 34. hypoventilation
_____ 35. severe diarrhea

**Multiple Choice** (Answers on page A-27)

_____ 36. Which of the following disorders is characterized by a deficiency of vasopressin?
      a. diabetes mellitus
      b. hemorrhage
      c. hypotonicity
      d. colloidal isotonicity
      e. diabetes insipidus

_____ 37. Which of the following disorders or symptoms can be caused by hypotonicity?
      a. diabetes insipidus
      b. hypertension and edema
      c. hemorrhage
      d. diarrhea
      e. diabetes mellitus

_____ 38. Which of the following best describes cerebrospinal fluid?
      a. tissue fluid
      b. ICF
      c. transcellular fluid
      d. lymph
      e. interstitial fluid

_____ 39. Which of the avenues for salt output eliminates the most salt per day?
      a. tears
      b. feces
      c. hemorrhage
      d. sweat
      e. controlled excretion in urine

_____ 40. Which is the primary cation found in the ECF?
      a. bicarbonate
      b. phosphate
      c. sodium
      d. chloride
      e. potassium

_____ 41. Which is the primary cation found in the ICF?
      a. bicarbonate
      b. sodium
      c. chloride
      d. phosphate
      e. potassium

_____ 42. Which of the following organs is responsible for long-term regulation of blood pressure?
      a. lungs
      b. heart
      c. spleen
      d. kidneys
      e. intestines

_____ 43. Which fluid compartment has the greatest percentage of body fluid?
      a. lymph
      b. plasma
      c. ICF
      d. ECF
      e. transcellular fluid

_____ 44. Which fluid compartment represents the greatest percentage of body fluid within the ECF?
      a. lymph
      b. interstitial fluid
      c. synovial fluid
      d. plasma
      e. cerebrospinal fluid

_____ 45. When the plasma volume is too large, to which fluid compartment is the excess fluid shifted?
      a. lymph
      b. interstitial fluid
      c. transcellular fluid
      d. cerebrospinal fluid
      e. none of the above

**Points to Ponder**

1. As you shop in your favorite grocery store you may select a meat product that has been "sugar cured". How does this process protect the meat and make it safe for you to eat?

2. At present one of the makers of a common antacid boasts that their product uses calcium instead of sodium and this is better. Do you find any truth in this advertisement? Explain.

3. Most Americans are aware of the caffeine in those popular drinks; coffee, tea, and most soft drinks. How are the other ingredients related to this chapter? Are these drinks diuretics? Are these beverages acidic or basic? How does your body handle these fluids?

**Clinical Perspective**

1. The patient has lost a lot of blood due to the traumatic injuries received in an auto accident. Intravenous fluids are administered. Start with the hemorrhage and explain the sequence of events with reference to this chapter.

**Experiment of the Day**

1. Equipment:
   - 3 glasses
   - 3 dried prunes
   - sugar
   - salt

Place water in all three glasses. Place several spoons of salt into one glass and several spoons of sugar into still another. The third glass has plain water. Drop a prune into each glass. Note time. Explain the results.

# DIGESTIVE SYSTEM 13

## CHAPTER OVERVIEW

Cells need nutrients, vitamins, and minerals to survive.
The digestive system meets these cellular needs by removing the substances from the external environment and presenting them to the cells in a form that the cells can utilize. The food systematically passes through the mouth, pharynx, esophagus, stomach, small intestine, large intestine, and anus. The digestive tract receives enzymes and other secretions from the salivary glands, the exocrine pancreas, and the liver, and from exocrine cells in its epithelial lining. During this trip, carbohydrates, proteins and fats are broken down by various enzymes and other substances to yield monosaccharides, amino acids, and monoglycerides respectively. These smaller nutrient molecules are then absorbed into the small intestine and pass to the liver or lymphatic system. The liver performs the final modification on the nutrient molecules. Undigested or unutilized material along with bacteria are passed to the outside through the anus. Nutrition prepared by the digestive system is another important step in homeostasis. The digestive system is under the control of the central nervous system and several hormones.

## CHAPTER OUTLINE

### Introduction (Text page 419)

The primary function of the digestive system is to transfer nutrients, water, and electrolytes from the food we eat into the body's internal environment.

### The digestive system performs four basic digestive processes (page 419).

There are four basic digestive processes: motility, secretion, digestion, and absorption.

Propulsive movements push the contents forward through the digestive tract at varying speeds.

Mixing movements serve a twofold function.

First, by mixing food with the digestive juices, these movements promote digestion of the food.

Second, they facilitate absorption by exposing all portions of the intestinal contents to the absorbing surfaces of the digestive tract.

Each digestive secretion consists of water, electrolytes, and specific organic constituents that are important in the digestive process, such as enzymes, bile salts, or mucus.

Digestion refers to the breaking-down process whereby the structurally complex foodstuffs of the diet are converted to smaller absorbable units by the enzymes produced within the digestive system.

Digestion is accomplished by enzymatic hydrolysis.

Through the process of absorption, the small absorbable units that result from digestion, along with water, vitamins, and electrolytes, are transferred from the digestive tract lumen into the blood or lymph.

## The digestive tract and accessory digestive organs make up the digestive system (page 421).

The accessory digestive organs include the salivary glands, the exocrine pancreas, and the biliary system (liver and gallbladder).

The digestive tract includes the following organs: mouth; pharynx; esophagus; stomach; small intestine; large intestine; and anus.

A cross section of the digestive tube reveals four tissue layers: mucosa, submucosa, muscularis externa, and serosa.

## Regulation of digestive function is complex and synergistic (page 424).

Four factors are involved in the regulation of digestive system function: (1) autonomous smooth muscle function, (2) intrinsic nerve plexuses, (3) extrinsic nerves, and (4) gastrointestinal hormones.

The prominent type of self-induced electrical activity in digestive smooth muscle is slow-wave potentials, alternately referred to as the digestive tracts basic electrical rhythm.

The second factor involved in the regulation of digestive-tract function is the intrinsic nerve plexuses.

Two major networks of nerves form the plexus of the digestive tract: the myenteric plexus and the submucous plexus.

The two plexuses are often termed the enteric nervous system.

The extrinsic nerves are from both branches of the autonomic system.

The sympathetic system tends to inhibit or slow down digestive tract contraction and secretion.

The parasympathetic nervous system dominates in quiet relaxed situations when general maintenance types of activities such as digestion can proceed optimally.

Parasympathetic nerve fibers which arrive primarily by way of the vegas nerve, tend to increase smooth muscle motility and promote secretion of digestive enzymes and hormones.

Tucked within the mucosa of certain regions of the digestive tract are endocrine gland cells that release hormones into the blood upon appropriate stimulation.

Gastrointestinal hormones are carried to other areas of the digestive tract where they exert either excitatory or inhibitory influences on smooth muscle and exocrine gland cells.

Gastrointestinal hormones are released primarily in response to specific local changes in the luminal contents, acting either directly on the endocrine gland cells or indirectly through the intrinsic plexuses or extrinsic autonomic nerves.

### Receptor activation alters digestive activity through neural reflexes and hormonal pathways (page 425).

The wall of the digestive tract contains three different types of sensory receptors that respond to local chemical or mechanical changes in the digestive tract: (1) chemo-receptors; (2) mechanoreceptors; and (3) osmoreceptors.

The effector cells include smooth muscle, exocrine gland cells and endocrine gland cells.

## Mouth (Text page 425)

### The oral cavity is the entrance to the digestive tract (page 425).

Entry to the digestive tract is through the mouth or oral cavity.

The palate separates the mouth from the nasal passages.

Toward the front of the mouth, the palate is made of bone, forming what is known as the hard palate.

The palate toward the rear of the mouth is called the soft palate.

The tongue is composed of voluntarily controlled skeletal muscle.

The pharynx acts as a common passageway for both the digestive system and the respiratory system.

### The teeth are responsible for chewing, which breaks up food, mixes it with saliva, and stimulates digestive secretions (page 426).

The first step in the digestive process is mastication.

The purposes of chewing are: (1) to grind and break food up into smaller pieces to facilitate swallowing; (2) to mix food with saliva; and (3) to stimulate taste buds.

### Saliva begins carbohydrate digestion but plays more important roles in oral hygiene and in facilitating speech (page 426).

Saliva is produced by three major pairs of salivary glands.

Saliva is composed of water, proteins and electrolytes.

The most important salivary proteins--amylase, mucus, and lysozyme--contribute to the functions of saliva as follows: (1) salivary amylase breaks polysaccharides down into disaccharides; (2) saliva facilitates swallowing by providing mucus; (3) saliva exerts some antibacterial action by means of lysozyme, an enzyme that lyses or destroys certain bacteria; (4) saliva serves as a solvent for molecules that stimulate the taste buds; (5) saliva aids speech by facilitating movements of the lips and tongue; (6) saliva plays an important role in oral hygiene; and (7) saliva buffers the acids in food as well as acids produced by bacteria in the mouth, thereby helping to prevent dental caries.

### The continuous low level of salivary secretion can be increased by simple and conditioned reflexes (page 427).

The simple, or unconditioned, salivary reflex occurs when chemoreceptors and pressure receptors within the oral cavity respond to the presence of food.

These receptors initiate impulses in afferent nerve fibers that carry the information to the salivary center located in the medulla of the brain stem.

The salivary center sends impulses via the extrinsic autonomic nerves to the salivary glands to promote increased salivation.

With the acquired, or conditioned, salivary reflex, salivation occurs without oral stimulation.

Just thinking about, seeing, smelling, or hearing the preparations of pleasant food initiates salivation through this reflex.

Inputs that arise outside the mouth and are mentally associated with the pleasure of eating act through the cerebral cortex to stimulate the medullary salivary center.

Parasympathetic stimulation, which exerts the dominant role in salivary secretion, produces a prompt and abundant flow of watery saliva that is rich in enzymes.

Sympathetic stimulation produces a much smaller volume of thick saliva that is rich in mucus.

## Digestion in the mouth is minimal, and no absorption of nutrients occurs (page 428).

Importantly, some therapeutic agents can be absorbed by the oral mucosa, a prime example being a vasodilator drug, nitroglycerin, which is used by certain cardiac patients to relieve anginal attacks associated with myocardial ischemia.

# Pharynx and Esophagus (Text page 428)

## Swallowing is a sequentially programmed all-or-none reflex (page 428).

The pressure of the bolus in the pharynx stimulates pharyngeal pressure receptors, which send afferent impulses to the swallowing center located in the medulla.

Swallowing is an example of a sequentially programmed all-or-none reflex in which multiple responses are triggered in a specific timed sequence.

Swallowing is initiated voluntarily, but once it is initiated, it cannot be stopped.

## During the oropharyngeal stage of swallowing, food is directed into the esophagus and is prevented from entering the wrong passageways (page 428).

Contraction of laryngeal muscles aligns the vocal cords in tight apposition to each other, thus sealing the glottis entrance.

The bolus tilts the epiglottis backward down over the closed glottis.

Because the respiratory passages are temporarily sealed off during swallowing, respiration is briefly inhibited.

Pharyngeal muscles contract to force the bolus into the esophagus.

## The esophagus is guarded by sphincters at both ends (page 428).

The upper esophageal sphincter is the pharyngoesophageal sphincter and the lower sphincter is the gastroesophageal sphincter.

## Peristaltic waves push food through the esophagus (page 429).

The swallowing center initiates a primary peristaltic wave that sweeps from the beginning to the end of the esophagus, forcing the bolus ahead of if through the esophagus to the stomach.

If a bolus fails to be carried along to the stomach, a second, more forceful peristaltic wave is mediated by the intrinsic nerve plexus.

The gastroesophageal sphincter prevents reflux of gastric contents (page 430).
>    If gastric contents do flow back into the esophagus, the acidity of these contents irritates the esophagus, causing the esophageal discomfort known as heartburn.

Esophageal secretion is entirely protective (page 430).
>    Esophageal secretion is entirely mucus.
>    Mucus provides lubrication for passage of food and protects the esophageal wall from acid and enzymes in the gastric juice if gastric reflux should occur.

# Stomach (Text page 430)

The stomach stores food and begins protein digestion (page 430).
>    The fundus is the portion of the stomach that lies above the esophageal opening.
>    The main part of the stomach is the body.
>    The lower portion is called the antrum.
>    The terminal portion of the stomach consists of the pyloric sphincter.
>    It is important that the stomach store food and meter it into the duodenum at a rate that does not exceed the small intestines capacities.
>    A second function of the stomach is to secrete hydrochloric acid and enzymes that begin protein digestion.

Stomach motility is complex and subject to multiple regulatory inputs (page 430).
>    There are four aspects to gastric motility: (1) gastric filling, (2) gastric storage, (3) gastric mixing, and (4) gastric emptying.
>    The reflex relaxation of the stomach as it receives food is called receptive relaxation.
>    Food emptied into the stomach from the esophagus is stored in the relatively quiet body without being mixed.
>    Food is gradually fed from the body into the antrum, where mixing does take place.
>    The strong antral peristaltic contractions are responsible for the mixing of food with gastric secretions to produce chyme.
>    The antral peristaltic contractions provide the driving force for gastric emptying.
>    The main gastric factor that influences the strength of contraction is the amount of chyme in the stomach.
>    Distention of the stomach triggers increased gastric motility through a direct effect of stretch on the smooth muscle as well as through involvement of the intrinsic plexuses, the vagus nerve, and the stomach hormone gastrin.
>    Factors in the duodenum are of primary importance in controlling the rate of gastric emptying.
>    The four most important duodenal factors that influence gastric emptying are fat, acid, hypertonicity and distention.

Emotions can influence gastric motility (page 433).
>    Sadness and fear generally tend to decrease motility, whereas anger and aggression tend to increase it.
>    Intense pain from any part of the body tends to inhibit motility.

The body of the stomach does not actively participate in the act of vomiting (page 434).

> Vomiting or emesis, is the forceful expulsion of gastric contents out through the mouth.
>
> The major force for expulsion comes from contraction of the respiratory muscles--- namely, the diaphragm and the abdominal muscles.
>
> The complex act of vomiting is coordinated by a vomiting center in the medulla.
>
> The causes of vomiting include the following: (1) tactile stimulation of the back of the throat; (2) irritation or distension of the stomach or duodenum; (3) elevated intracranial pressure; (4) rotation or acceleration of the head producing dizziness; (5) chemical agents, emetics; and (6) psychological vomiting induced by emotional factors.

Gastric pits are the source of gastric digestive secretions (page 434).

> The cells responsible for gastric secretion are located in the lining of the stomach, the gastric mucosa, which is divided into two distinct areas; (1) the oxyntic mucosa, which lines the body and the fundus, and (2) the pyloric gland area, which lines the antrum.
>
> The mucosal gland cells are found in gastric pits.
>
> Three types of secretory cells are found in the walls of the pits in the oxyntic mucosa.
>
> The mucous neck cells secrete a thin, watery mucus.
>
> Chief cells secrete the enzyme precursor pepsinogen and parietal cells secrete HCl and intrinsic factor.
>
> The gastric pits of the pyloric gland area secrete mucus and pepsinogen.
>
> No acid is secreted in the pyloric gland area.
>
> More importantly, endocrine cells in the pyloric gland area secrete the hormone gastrin into the blood.
>
> Parietal cells actively secrete HCl.
>
> The pH of the luminal contents falls as low as 2 as a result of this HCl secretion.
>
> Hydrochloric acid (1) activates the enzyme precursor pepsinogen to an active enzyme, pepsin, and provides an acid medium that is optimal for pepsin activity, (2) aids in the breakdown of connective tissue and muscle fibers, and (3) along with salivary lysozyme, kills most of the microorganisms ingested with the food.
>
> The major digestive constituent of gastric secretion is pepsinogen.
>
> Pepsinogen is stored in the chief cell's cytoplasm within secretory vesicles known as zymogen granules.
>
> Hydrochloric acid cleaves off a small fragment of the pepsinogen molecule, converting it to the active form of the enzyme, pepsin.
>
> Pepsin initiates protein digestion by splitting proteins to yield peptide fragments.
>
> The surface of the gastric mucosa is covered by a layer of mucus.
>
> Mucus protects the gastric mucosa against mechanical injury.
>
> Mucus helps protect the stomach wall from self-digestion.
>
> Being alkaline, mucus helps protect against acid injury by neutralizing HCl in the vicinity of the gastric lining.
>
> Intrinsic factor is important in the absorption of certain vitamins.
>
> Gastrin stimulates the parietal and chief cells.

The most potent stimulus for increased gastric secretion is protein in the stomach (page 436).

> The cephalic phase of gastric secretion refers to the increased secretion of HCl and pepsinogen even before the food reaches the stomach.

Thinking about, tasting, smelling, chewing, and swallowing food increases gastric secretion by means of vagal nerve activity.

Vagal stimulation of the intrinsic plexuses promotes increased secretion of HCl and pepsinogen by the secretory cells.

Vagal stimulation of the pyloric gland cells causes a release of gastrin.

The gastric phase of gastric secretion occurs when food actually reaches the stomach.

Stimuli acting in the stomach---namely, proteins; distention; caffeine; or alcohol---increase gastric secretion by means of overlapping efferent pathways.

Protein initiates short local reflexes in the intrinsic nerve plexuses to stimulate the secretory cells.

Protein stimulates vagal fibers to the stomach.

Vagal activity further enhances intrinsic nerve stimulation of the secretory cells and triggers the release of gastrin.

Protein directly stimulates the release of gastrin.

## Gastric secretion gradually decreases as food empties from the stomach into the intestine (page 437).

As the meal is gradually emptied into the duodenum, the major stimulus for enhanced gastric secretion is withdrawn.

After foods leaves the stomach the pH falls very low.

Gastric secretion is inhibited.

The same stimuli that inhibit gastric motility inhibit gastric secretion.

## The stomach lining is protected from gastric secretions by the gastric mucosal barrier (page 437).

The properties of the gastric mucosa that enable the stomach to contain acid without injuring itself constitute the gastric mucosal barrier.

The barrier is occasionally broken so that the gastric wall is injured by its acidic and enzymatic contents.

When this occurs, an erosion, or peptic ulcer, of the stomach wall exists.

Two of the most serious consequences of ulcers are (1) hemorrhage resulting from damage of submucosal capillaries and (2) perforation of the stomach wall due to complete erosion through the wall caused by HCl and pepsin actin.

## Carbohydrate digestion continues in the body of the stomach, whereas protein digestion begins in the antrum (page 439).

Carbohydrate digestion continues under the influence of salivary amylase.

## The stomach absorbs alcohol and aspirin but no food (page 439).

No food or water is absorbed into the blood from the stomach mucosa.

Even though none of the ingested food is absorbed from the stomach, two noteworthy nonnutrient substances are absorbed directly by the stomach----ethyl alcohol and aspirin.

## Pancreatic and Biliary Secretions (Text page 439)

### The pancreas is a mixture of exocrine and endocrine tissue (page 439).

The pancreas is a mixed gland that contains both exocrine and endocrine tissue.

The predominant exocrine portion consists of grapelike clusters of secretory cells that form sacs known as acini.

The smaller endocrine portion consists of isolated islands of endocrine tissue, the islets of Langerhans.

### The exocrine pancreas secretes digestive enzymes and an aqueous alkaline fluid (page 440).

The exocrine pancreas secretes a pancreatic juice consisting of two components--- a potent enzymatic secretion and an aqueous alkaline secretion that is rich in sodium bicarbonate.

The three types of pancreatic enzymes are (1) proteolytic enzymes; (2) pancreatic amylase; and (3) pancreatic lipase.

The three major proteolytic enzymes secreted by the pancreas are trypsinogen, chymo-trypsinogen, and procarboxypeptidase.

When trypsinogen is secreted into the duodenal lumen, it is activated to its active enzyme form, trypsin, by enterokinase.

Chymotrypsinogen and procarboxypeptidase are converted to their active forms, chymotrypsin and carboxypeptidase within the duodenal lumen.

Each of the proteolytic enzymes attacks different peptide linkages.

Pancreatic amylase plays an important role in carbohydrate digestion by converting polysaccharides into disaccharides.

Pancreatic lipase is extremely important because it is the only enzyme secreted through-out the entire digestive system that can accomplish digestion of fat.

Pancreatic enzymes function best in a neutral or slightly alkaline environment.

The alkaline fluid secreted by the pancreas into the duodenal lumen serves the important function of neutralizing the acidic chyme.

### Pancreatic exocrine secretion is hormonally regulated to maintain neutrality of the duodenal contents and to optimize digestion (page 441).

A small amount of parasympathetically induced pancreatic secretion occurs during the cephalic phase of digestion.

The release of two major enterogastrones, secretin and cholecystokinin, in response to chyme in the duodenum plays the central role in the control of pancreatic secretion.

The primary stimulus specifically for secretion release is acid in the duodenum.

Secretion is carried by the blood to the pancreas, where it stimulates the duct cells to markedly increase their secretion of a sodium bicarbonate--rich aqueous fluid into the duodenum.

This mechanism provides a control system for maintaining neutrality of the chyme in the intestine.

Cholecystokinin is important in the regulation of pancreatic enzyme secretion.

The circulatory system transport cholecystokinin to the pancreas, where it stimulates the pancreatic acinar cells to increase secretion of lipase and the proteolytic enzymes.

The liver performs various important functions, including bile production (page 442).

> The biliary system includes the liver, the gallbladder, and associated ducts.
>
> The liver's importance to the digestive system is its secretion of bile, but it performs a wide variety of other functions, including the following: (1) metabolic processing of the major categories of nutrients; (2) detoxification or degradation of body wastes and hormones as well as drugs and other foreign compounds; (3) synthesis of plasma proteins; (4) storage of glycogen, fats, copper, iron, and many vitamins; (5) activation of vitamin D; (6) removal of bacteria and worn-out red blood cells; and (7) excretion of cholesterol and bilirubin.
>
> Venous blood enters the liver by means of a unique and complex vascular connection between the digestive tract and the liver that is known as the hepatic portal system.
>
> The veins draining the digestive tract do not directly join the inferior vena cava.
>
> The veins from the stomach and intestine enter the hepatic portal vein, which carries the products absorbed from the digestive tract directly to the liver for processing, storage, or detoxification.

The liver lobules are delineated by vascular and bile channels (page 443).

> Hepatocytes continuously secrete bile into thin channels, the bile canaliculi, which carry the bile to a bile duct at the periphery of the lobule.

Bile is secreted by the liver and is diverted to the gallbladder between meals (page 443).

> The opening of the bile duct into the duodenum is guarded by the sphincter of Oddi.
>
> When the sphincter is closed, the bile is diverted back up into the gallbladder.

Bile salts are recycled through the enterohepatic circulation (page 443).

> Bile consists of an aqueous alkaline fluid as well as several organic constituents, including bile salts, cholesterol, lecithin, and bilirubin.
>
> Bile salts are derivatives of cholesterol.

Bile salts aid fat digestion and absorption through their detergent action and micellar formation, respectively (page 444).

> Detergent actions refers to the bile salt's ability to convert large fat globules into a liquid emulsion that consists of many small fat droplets suspended in the aqueous chyme, thus increasing the surface area available for attack by pancreatic lipase.
>
> Bile salts, along with cholesterol and lecithin, play an important role in facilitating fat absorption through micellar formation.
>
> Micelles, being water soluble by virtue of their hydrophobic shells, can dissolve water-insoluble substances in their lipid-soluble cores.

Bilirubin is a waste product secreted in the bile (page 446).

> Bilirubin does not play a role in digestion but is one of the few waste products excreted in the bile.
>
> Bilirubin is the end product resulting from the degradation of the heme portion of hemoglobin.
>
> Bilirubin is a yellow pigment that gives bile its yellow color.

If bilirubin is formed more rapidly than it is excreted, it accumulates in the body and causes jaundice.

Jaundice can be brought about in three different ways: (1) prehepatic or hemolytic jaundice, (2) hepatic jaundice; and (3) posthepatic or obstructive jaundice.

## Bile salts are the most potent stimulus for increased bile secretion (page 446).

Any substance that increases bile secretion is called a choleretic.

The most potent choleretic is bile salts themselves.

Secretin stimulates a aqueous alkaline bile secretion by the liver.

Vagal stimulation of the liver plays a minor role in bile secretion during the cephalic phase of digestion.

## The gallbladder stores and concentrates bile between meals and empties during meals (page 446).

The presence of food in the duodenal lumen triggers the release of cholecystokinin.

Cholecystokinin stimulates contraction of the gallbladder and relaxation of the sphincter of Oddi, so bile is discharged into the duodenum.

# Small Intestine (Text page 447)

The small intestine is the site at which most digestion and absorption take place.

The small intestine is arbitrarily divided into three segments: (1) the duodenum; (2) the jejunum; and (3) the ileum.

## Segmentation contractions mix and slowly propel the chyme (page 447).

Segmentation consists of oscillating, ringlike contractions of the circular smooth muscle along the length of the intestine.

The contractile rings occur every few centimeters, dividing the small intestine into segments like a chain of sausages.

After a brief period of time, the contracted segments relax, and ringlike contractions appear in the previously relaxed segments.

Parasympathetic stimulation enhances segmentation.

Sympathetic stimulation depresses segmental activity.

## The migrating motility complex sweeps the intestine clean between meals (page 448).

When most of the meal has been absorbed, segmentation contractions cease and are replaced between meals by the migrating motility complex.

## The ileocecal juncture prevents contamination of the small intestine by colonic bacteria (page 448).

The anatomical arrangement is such that valvelike folds of tissue protrude from the ileum into the lumen of the cecum.

This ileocecal valve is easily pushed open, but the folds of tissue are forcibly closed ` when the cecal contents attempt to move backward.

The smooth muscle within the last several centimeters of the ileal wall is thickened, forming a sphincter that is under neural and hormonal control.

Small-intestine secretions do not contain any digestive enzymes (page 448).

> The exocrine glands located in the small-intestine mucosa secrete into the lumen an aqueous salt and mucus solution known as the succus entericus.
>
> The mucus in the secretion provides protection and lubrication.

Digestion in the small intestine lumen is accomplished by pancreatic enzymes, whereas the small intestine enzymes act intracellularly (page 449).

> Digestion within the small intestine lumen is accomplished by the pancreatic enzymes, with fat digestion being enhanced by bile secretion.
>
> Fat digestion is completed within the small intestine lumen.
>
> Hairlike projections on the luminal surface of the small intestine epithelial cells form the brush border, which contains three different categories of enzymes: (1) enterokinase (2) the disaccharidases; and (3) the aminopeptidases.

The small intestine is remarkably well adapted for its primary role in absorption (page 450).

> Most absorption occurs in the duodenum and jejunum.
>
> The small intestine has an abundant reserve absorptive capacity.
>
> The following special modifications of the small intestine mucosa greatly increase the surface area available for absorption: (1) the inner surface of the small intestine is thrown into circular folds; (2) projecting from this folded surface are microscopic fingerlike projections known as villi; and (3) even smaller hairlike projections known as microvilli arise from the luminal surface of the epithelial cells.

The mucosal lining experiences rapid turnover (page 451).

> The epithelial cells lining the small intestine slough off and are replaced at a rapid rate as a result of high mitotic activity in the crypts of Lieberkuhn.
>
> The old cells that are sloughed off into the lumen are digested, with the cell constituents being absorbed into the blood and reclaimed for synthesis of new cells, among other things.

Special mechanisms facilitate absorption of most nutrients (page 451).

> Sodium may be absorbed and actively.
>
> Most water absorption in the digestive tract depends on the active carrier that pumps sodium.
>
> Dietary carbohydrate is presented to the small intestine for absorption mainly in the forms of the disaccharides maltose, sucrose, and lactose.
>
> Glucose and galactose are both absorbed by secondary active transport.
>
> The operation of these cotransport carriers depends on the sodium concentration gradient established by the energy-consuming basolateral sodium-potassium pump.
>
> Glucose leaves the cell down its own concentration gradient to enter the blood.
>
> Not only are ingested proteins digested and absorbed, but endogenous proteins are digested and absorbed as well.
>
> Amino acids are absorbed across the intestinal cells by secondary active transport.
>
> Fat absorption is quite different from carbohydrate and protein absorption because of the insolubility of fat in water presents a special problem.
>
> Biliary components facilitate absorption of these fatty end products through formation of micelles.

Once the micelles reach the luminal membranes of the epithelial cells, the mono-glycerides and free fatty acids passively diffuse from the micelles through the lipid component of the epithelial cell membranes to enter the interior of these cells.
Once within the interior of epithelial cells, the monoglycerides and free fatty acids are resynthesized into triglycerides.
The large, coated fat droplets, known as chylomicrons, enter the central lacteals rather than the capillaries because of the structural differences between these two vessels.
Fat can be absorbed into the lymphatics but not directly into the blood.
Fat absorption is a passive process.
Water-soluble vitamins are primarily absorbed passively with water, whereas fat-soluble vitamins are carried in the micelles and absorbed passively with the end products of fat digestion.

## Most absorbed nutrients immediately pass through the liver for processing (page 454).

Anything absorbed into the digestive capillaries first must pass through the hepatic biochemical factory before entering the general circulation.
Fat is picked up by the central lacteal and enters the lymphatic system instead.
The absorbed fat is carried by systemic circulation to the liver and to other tissues of the body.

## Extensive absorption by the small intestine keeps pace with secretion (page 454).

The digestive juices are not lost from the body.
After the constituents of the juices are secreted into the digestive tract lumen and perform their function, they are returned to the plasma.

## Diarrhea results in loss of fluid and electrolytes (page 454).

The most common cause of diarrhea is excessive intestinal motility, which arises either from local irritation of the gut wall by bacteria or viral infection of the intestine or from emotional stress.

# Large Intestine (Text page 454)

## The large intestine is primarily a drying and storage organ (page 454).

The colon extracts more water and salt from the contents.
What remains to be eliminated is known as feces.

## Haustral contractions slowly shuffle the colonic contents back and forth while mass movements propel colonic contents long distances (page 455).

The colon's primarily method of motility is Haustral contractions initiated by the autonomous rhymicity of colonic smooth-muscle cells.
Haustral contractions are largely controlled by locally mediated reflexes involving the intrinsic plexuses.
The gastrocolic reflex is mediated from the stomach to the colon by gastrin and by the extrinsic autonomic nerves.
The gastroileal reflex moves the remaining small intestine contents into the large intestine, and the gastrocolic reflex pushes the colonic contents into the rectum, triggering the defecation reflex .

Feces are eliminated by the defecation reflex (page 456).

When mass movements of the colon move fecal material into the rectum, the resultant distention of the rectum stimulates stretching receptors in the rectal wall, thus initiating the defecation reflex.

The defecation reflex causes the internal anal sphincter to relax and the rectum and sigmoid colon to contract more vigorously.

If the external anal sphincter is also relaxed, defecation occurs.

Constipation occurs when the feces become too dry (page 456).

If defecation is delayed too long, constipation may result.

When the colonic contents are retained for longer periods of time than normal, more than the usual amount of water is absorbed.

Large intestine secretion is protective in nature (page 456).

Colonic secretion consists of an alkaline mucus solution, whose function is to protect the large-intestine mucosa from mechanical and chemical injury.

The large intestine absorbs salt and water, converting the luminal contents into feces (page 456).

The colon normally absorbs some salt and water.

Intestinal gases are absorbed or expelled (page 457).

Flatus is derived primarily from two sources: (1) swallowed air; and (2) gas produced by bacterial fermentation in the colon.

Some foods, such as beans, contain types of carbohydrates for which humans lack digestive enzymes.

These fermentable carbohydrates enter the colon, where they are attacked by gas producing bacteria.

# Overview of Gastrointestinal Hormones (Text page 457)

Chyme in the stomach stimulates the release of gastrin, which acts on parietal and chief cells to increase secretion of HCl and pepsinogen.

Gastrin enhances gastric motility, stimulates ileal motility, relaxes the ileocecal sphincter, and induces mass movements in the colon.

Secretion performs four major interrelated functions: (1) it inhibits gastric emptying to prevent further acid from entering the duodenum until the acid that is present is neutralized; (2) it inhibits gastric secretion to reduce the amount of acid being produced; (3) it stimulates the pancreatic duct cells to produce a large volume of aqueous sodium bicarbonate secretion, which is emptied into the duodenum to neutralize the acid; and (4) it stimulates secretion by the liver of a sodium-bicarbonate-rich bile, which likewise is emptied into the duodenum to assist in the neutralization process.

Cholecystokinin performs several important interrelated functions: (1) it inhibits gastric motility and secretion, thereby allowing adequate time for the nutrients already in the duodenum to be digested and absorbed; (2) it stimulates the pancreatic acinar cells to increase secretion of pancreatic enzymes, which continue the digestion of these nutrients in the duodenum; and (3) it causes contraction of the gallbladder and relaxation of the sphincter of Oddi so that bile is emptied into the duodenum to aid fat digestion and absorption.

Cholecystokinin is an important regulator of food intake.
Cholecystokinin plays a key role in satiety, the sensation of having had enough to eat.

## KEY TERMS

Bilirubin-p446
Bolus-p428
Carboxypeptidase-p440
Cholecystokinin-p432
Choleretic-p446
Chymotrypsin-p440
Chymotrypsinogen-p440
Deglutition-p428
Emetics-p434
Enterokinase-p428

Gastric inhibitory peptide-p432
Gastrin-p436
Haustral-p455
Hepatic portal system-p442
Hydrolysis-p421
Jaundice-p446
Micelle-p444
Microvilli-p450
Pepsin-p436
Pepsinogen-p436

Peristalsis-p429
Procarboxypeptidase-p440
Salivary amylase-p426
Secretin-p432
Succus entericus-p448
Trypsin-p440
Trypsinogen-p440
Villikinin-p454
Xerostomia-p427
Zymogen granules-p436

## REVIEW EXERCISES

**True - False** (Answers on page A-28)

_____ 1. The simplest form of carbohydrate is the simple sugars or disaccharides.
_____ 2. Starch, glycogen, and cellulose are all disaccharides.
_____ 3. The end products of fat digestion are triglycerides and free fatty acids.
_____ 4. Digestion is accomplished by enzymatic hydrolysis.
_____ 5. The lamina propria is a layer in the submucosa of the digestive tract.
_____ 6. Meissner's and Haustra's plexuses together are termed the enteric nervous system.
_____ 7. The palate separates the mouth from the nasal passages.
_____ 8. The tongue is composed of voluntarily controlled smooth muscle.
_____ 9. Bicarbonate buffers are found in saliva.
_____ 10. Salivary amylase breaks down polysaccharides.
_____ 11. The hard outer covering or the crown of a tooth is enamel.
_____ 12. Swallowing is known as deglutition.
_____ 13. Burping is called borborygmia.
_____ 14. The antrum is that portion of the stomach that lies above the esophageal opening.
_____ 15. The interior of the stomach is thrown into deep folds.
_____ 16. Gastrin influences the rate of gastric emptying.
_____ 17. Cholecystokinin is responsible for the contraction of the gallbladder.
_____ 18. Malocclusion is the forceful expulsion of gastric contents out through the mouth.
_____ 19. Chief cells secrete mucus.
_____ 20. Oxyntic cells is another name for mucous neck cells.

_____ 21. Hydrochloric acid activates pepsinogen into the active enzyme pepsin.
_____ 22. The major digestive constituent of gastric secretion is trypsinogen.
_____ 23. Peptic ulcers frequently occur when the gastric mucosal barrier is severely eroded or broken.
_____ 24. The stomach absorbs aspirin.
_____ 25. Water is absorbed from the stomach contents.
_____ 26. The exocrine portion of the pancreas is known as the islets of Langerhans.
_____ 27. Trypsinogen is activated to its active enzyme form trypsin by HCl.
_____ 28. Pancreatic juice is an aqueous acidic solution.
_____ 29. Cholecystokinin is important in the regulation of pancreatic enzyme secretion.
_____ 30. The principal clinical manifestation of pancreatic exocrine insufficiency is steatorrhea.
_____ 31. Macrophages of the liver are called Kupffer cells.
_____ 32. Bile is produced in the gallbladder.
_____ 33. Most absorption occurs in the duodenum and jejunum.
_____ 34. Small-intestine secretions do not contain any digestive enzymes.
_____ 35. Villi, the fingerlike projections in the small intestine, are also known as the brush border.
_____ 36. The acid-base balance of the body is not altered by the digestive processes.
_____ 37. The colon's primary method of motility is Haustral contractions.
_____ 38. Cholecystokinin inhibits gastric motility.
_____ 39. Secretin inhibits gastric secretion.

**Fill in the Blank** (Answers on page A-29)
40. The exocrine glands located in the small intestine mucosa secrete into the lumen an aqueous salt and mucus solution known as the _____.
41. _____ activates the pancreatic enzyme trypsinogen.
42. _____ hydrolyze the small peptide fragments into their amino acids components, completing protein digestion.
43. Projecting from the small intestines folded surface are microscopic fingerlike projections known as _____.
44. _____ may be caused by damage to or reduction of the surface area of the small intestine.
45. Dipping down into the mucosal surface between the villi are shallow invaginations known as the _____.
46. The first step in the digestive process is _____.
47. Saliva is composed of about _____ $H_2O$ and _____ protein and electrolytes.
48. _____ is a condition characterized by decreased salivary secretions.
49. With _____ salivation occurs without oral stimulation.
50. The _____ separates the mouth from the nasal passages.

51. Some therapeutic agents can be absorbed by the oral mucosa, an example is

    _____.

52. A _____ is a ball of food.

53. The _____ consists of moving the bolus from the mouth through the pharynx and into the esophagus.

54. The lower esophageal sphincter is the _____.

55. _____ refers to ringlike contractions of the circular smooth muscle that move progressively forward with a stripping motion.

56. _____ do not involve the swallowing center.

57. The small fingerlike projection at the bottom of the cecum is the _____.

58. The colon's primary method of motility is _____.

59. The large intestine consists of the _____, _____, _____ and

    _____.

60. _____ plays an important role in inhibiting gastric motility and secretion in response to fat, acid, hypertonicity and distention in the duodenum.

61. The _____ is the portion of the stomach that lies above the esophageal opening.

62. The lower portion of the stomach is called the _____.

63. The mucosal gland cells are found in _____.

64. Endocrine cells in the PGA secrete the hormone _____ into the blood.

65. Pepsinogen is stored in the chief cell's cytoplasm within secretory vesicles known as

    _____.

66. _____ phase of gastric secretion occurs when food actually reaches the stomach.

67. A _____ occurs when the gastric wall is injured by its acidic and enzymatic contents.

68. _____ is a condition characteristic of excessive undigested fat in the feces.

69. The _____ is the largest and most important metabolic organ in the body.

70. The opening of the bile duct into the duodenum is guarded by the _____.

71. _____ are derivatives of cholesterol.

72. Any substance that increases bile secretion by the liver is called a _____.

73. The presence of food in the duodenal lumen triggers the release of _____.

74. Digestion is accomplished by enzymatic _____.

**Matching:** Match the characteristics on the left to the condition on the right
(Answers on page A-29)

_____ 75. This condition is characterized by diminished
salivary secretion

_____ 76. This is the forceful expulsion of gastric contents
out through the mouth

_____ 77. This colinical manifestation is excessive
undigested fat in the feces

_____ 78. This condition is impaired absorption in the
intestines

_____ 79. This condition occurs when the gastric contents
flow back into the esophagus

_____ 80. In this condition the small intestine is abnormally
sensitive to gluten

_____ 81. This condition occurs when bilirubin accumulates
in the body

_____ 82. This fairly common disorder involves a deficiency
of lactose

_____ 83. This condition is characterized by an erosion of the
stomach wall

a. xerostomia
b. heartburn
c. emesis
d. peptic ulcer
e. steatorrhea
f. jaundice
g. lactose intolerance
h. malabsorption
i. gluten enteropathy

**Multiple Choice** (Answers on page A-29)

_____ 84. Which category of foodstuffs is characterized by having peptide bonds?
    a. starches
    b. carbohydrates
    c. proteins
    d. fats
    e. triglycerides

_____ 85. Where is the salivary center located within the brain?
    a. pons
    b. medulla
    c. cerebellum
    d. cerebrum
    e. hypothalamus

_____ 86. Which of the following is absorbed in the mouth?
    a. starch
    b. amino acids
    c. lipids
    d. ethyl alcohol
    e. nitroglycerin

_____ 87. Which of the hormones trigger gastric motility?
      a. gastrin
      b. pepsin
      c. trypsin
      d. gastrokinin
      e. villikinin

_____ 88. Which of the following pancreatic proteolytic enzymes are converted to their active form by trypsin?
      a. trypsinogen
      b. pepsinogen
      c. enterokinase
      d. pancreatic lipase
      e. procarboxypeptidase

_____ 89. Which of the following has specialized cells known as Kupffer cells?
      a. islets of Langerhans
      b. gallbladder
      c. liver
      d. salivary glands
      e. small intestines

_____ 90. The opening of the bile duct into the duodenum is guarded by this sphincter?
      a. esophageal sphincter
      b. pyloric sphincter
      c. ileocecal sphincter
      d. sphincter of Acinius
      e. sphincter of Oddi

_____ 91. Which of the following play an important role in facilitating fat absorption through micellar formation?
      a. apoenzymes
      b. bile salts
      c. cholecystokinin
      d. disaccharidases
      e. enterokinase

_____ 92. Approximately how much fluid is absorbed by the small intestine per day?
      a. 500 ml
      b. 1,000 ml
      c. 1,200 ml
      d. 9,000 ml
      e. 9,500 ml

**Points to Ponder**

1. How selective is the digestive tract in its absorption process? How would you explain the absorption of indigestible compounds such as inulin? What about the absorption of known carcinogens?
2. What happens to the pH in your mouth when you sleep with your mouth open?
3. After studying this chapter how do you feel about the disorder discussed in television advertisements known as irregularity?
4. In terms of digestion what is happening when food is cooked?

**Clinical Perspectives**

1. Are there serious consequences when a person strains to have a bowl movement while constipated?
2. How do you suppose the bile flows through the ducts of the biliary systems?

# ENERGY BALANCE AND
# TEMPERATURE REGULATION

**14**

## CHAPTER OVERVIEW

The thousands of chemical reactions within the body are frequently referred to as the metabolic processes of the organism. Metabolism requires energy and has an optimum temperature range. The survival of cells depends on certain intracellular chemical reactions. In maintaining homeostasis the body systems utilize chemical processes for the benefit of all cells. In this chapter, the balance of energy and the regulation of temperature necessary for homeostasis are considered.

The first law of thermodynamics states that energy cannot by created or destroyed but can be converted to another form. The second law of thermodynamics states that energy will be lost as energy is converted from one form to another. Thus, energy balance is interrelated to temperature production.

Inasmuch as energy is not supplied in a form that can be utilized directly, heat is produced as the body converts the energy from one form to another. To meet the energy needs of the organism and at the same time maintain an optimum temperature various body systems interact to provide an energy balance and adequate methods of thermo-regulation. The controls are the same i.e., nervous and endocrine. The phenomenon is homeostasis.

## CHAPTER OUTLINE

### Energy Balance (Text page 463)

Most food energy is ultimately converted into heat in the body (page 463).

According to the first law of thermodynamics, energy can neither by created nor destroyed.

The energy in nutrient molecules that is not used to energize work is transformed into thermal energy, or heat.

215

The energy in ingested foodstuffs constitutes energy input to the body.
Energy output or expenditure falls into two categories: external work and internal work.

## The metabolic rate is the rate of energy use (page 464).

The rate at which energy is expended by the body during both external and internal work is known as the metabolic rate.
The basal metabolic rate is a reflection of the body's "idling speed," or the minimal waking rate of internal energy expenditure.

## Energy input must equal energy output to maintain a neutral energy balance (page 465).

Since energy cannot be created or destroyed, energy input must equal energy output.
There are three possible states of energy balance: (1) neutral energy balance; (2) positive energy balance; and (3) negative energy balance.
To maintain a constant body weight, energy acquired through food intake must equal energy expenditure by the body.

## Food intake is controlled primarily by the hypothalamus, but the mechanisms involved are not fully understood (page 465).

Control of food intake is primarily a function of the hypothalamus.
The following factors are among those that have been hypothesized as contributing to the control of food intake: (1) the size of fat stores; (2) the extent of gastrointestinal distention; (3) the extent of glucose utilization; (4) the intensity of cell power production; (5) the level of cholecystokinin and (6) psychosocial influences.

## Persons suffering from anorexia nervosa have a pathological fear of gaining weight (page 468).

Chronic diseases such as renal failure, cancer, and tuberculosis are commonly accompanied by a lack of appetite.
Researchers have identified a protein called cachetin, which is released by the immune system and is believed to contribute to the development of emaciation.
Patients with anorexia nervosa, most commonly adolescent girls and young women, have a morbid fear of becoming fat.
Because they have an aversion to food, they eat very little and consequently lose considerable weight, perhaps even starving themselves to death.
Other characteristics of the condition include altered secretion of many hormones, absence of menstrual periods, and low body temperature.

# Temperature Regulation (Text page 468)

## Internal core temperature is homeostatically maintained at 100° F (page 468).

Humans are usually in environments cooler than their bodies, so they must constantly generate heat internally to maintain their body temperatures.
Most people suffer convulsion when the internal body temperature reaches about 106° F.
From a thermoregulatory viewpoint, the body may be conveniently viewed as a central core surrounded by an outer shell.
Skin temperatures may fluctuate between 68° and 104° F without damage.

Heat gain must balance heat loss to maintain a stable core temperature (page 469).

> Balance between heat input and output is frequently disturbed by (1) changes in internal heat production by exercise and (2) changes in the external environmental temperature.
>
> If the core temperature starts to fall, heat production is increased and heat loss is minimized so that normal temperature can be restored.
>
> If the temperature starts to rise above normal, it can be corrected by increasing heat loss and reducing heat production.

Heat exchange between the body and the environment takes place by radiation, conduction, convection, and evaporation (page 469).

> Heat always moves down its own concentration gradient, that is, down a thermal gradient from a warmer to a cooler region.
>
> Radiation is the emission of heat energy from the surface of a warm body in the form of electromagnetic waves.
>
> Conduction is the transfer of heat between objects of differing temperatures that are in direct contact with each other.
>
> Convection refers to the transfer of heat energy by air currents.
>
> When water evaporates from the skin surface, the heat required to transform water from a liquid to a gaseous state is absorbed from the skin, thereby cooling the body.
>
> Sweating is an active evaporative heat-loss process under sympathetic nervous control.
>
> Sweat is a dilute salt solution that is actively extruded to the surface of the skin by sweat glands dispersed all over the body.

The hypothalamus integrates a multitude of thermosensory inputs from both the core and the surface of the body (page 471).

> The hypothalamus serves as the body's thermostat.
>
> The hypothalamus must be continuously apprised of both the skin temperature and the core temperature by means of specialized temperature-sensitive receptors called thermoreceptors; peripheral thermoreceptors and central thermoreceptors.

Shivering is the primary involuntary means of increasing heat production (page 472).

> The hypothalamus takes advantage of the fact that increased skeletal-muscle activity generates more heat.
>
> Shivering consists of rhythmic, oscillating skeletal-muscle contractions that occur at a rapid rate of ten to twenty per second.
>
> Shivering is a very effective mechanism in increasing heat production.
>
> Nonshivering thermogenesis also plays a role in thermoregulation.
>
> Nonshivering thermogenesis is mediated by the hormones epinephrine and thyroid hormone, both of which increase heat production by stimulating fat metabolism.

The magnitude of heat loss can be adjusted by varying the flow of blood through the skin (page 472).

> Most of the flow of blood through the skin is for purposes of temperature regulation.
>
> Skin vasomotor responses are coordinated by the hypothalamus.

Increased sympathetic activity to the skin vessels produces heat-conserving vasoconstriction in response to cold exposure, whereas decreased sympathetic activity produces heat-losing vasodilation of the skin vessels in response to heat exposure.

The hypothalamus simultaneously coordinates heat production, heat loss and heat conservation mechanisms to regulate core temperature homeostatically (page 473).

In response to cold exposure, the posterior region of the hypothalamus directs increased heat production such as by shivering, while simultaneously decreasing heat loss by skin vasoconstriction and other measures.

During heat exposure the anterior part of the hypothalamus reduces heat production by decreasing skeletal-muscle activity and promotes increased heat loss by inducing skin vasodilation.

When even the maximal skin vasodilation is inadequate to rid the body of excess heat, sweating is brought into play to accomplish further heat loss through evaporation.

During a fever, the hypothalamic thermostat is "reset" at an elevated temperature (page 473).

Fever refers to an elevation in body temperature as a result of infection or inflammation.

In response to microbial invasion, white blood cells release endogenous pyrogen which acts on the hypothalamic thermoregulatory center to raise the setting of the thermostat. The hypothalamus now maintains the temperature at the new set level instead of maintaining normal body temperature.

Endogenous pyrogen raises the set point of the hypothalamic thermostat during fever production by triggering the local release of prostaglandins.

Aspirin reduces a fever by inhibiting the synthesis of prostaglandins.

## KEY TERMS

| | | |
|---|---|---|
| Anorexia nervosa-p468 | External work-p463 | Obesity-p466 |
| Appetite centers-p465 | Glucostatic theory-p466 | Radiation-p469 |
| Basal metabolic rate-p464 | Internal work-p463 | Satiety centers-p465 |
| Conduction-p469 | Ischymetric theory-p466 | Shivering-p472 |
| Convection-p469 | Kilocalorie-p464 | Sweating-p470 |
| Core temperature-p468 | Lipostatic theory-p466 | Thermal energy-p463 |
| Endogenous pyrogen-p473 | Metabolic rate -p464 | Thermal gradient-p469 |
| Evaporation-p470 | Nonshivering thermogenesis-p472 | Thermoreceptors-p472 |

## REVIEW EXERCISES

**True - False** (Answers on page A-30)

_____ 1. Internal work refers to the energy expended when skeletal muscles are contracted to move external objects.

_____ 2. Energy cannot be created, destroyed or converted from one form to another form.

_____ 3. The rate at which energy is expended by the body is known as the metabolic rate.

_____ 4. The appetite centers are located in the lateral regions of the thalamus.

_____ 5. The glucostatic theory proposes that satiety is signaled by increased glucose utilization.

_____ 6. Cachetin is released by the immune system.

_____ 7. Obesity is defined as excessive fat content in the adipose-tissue stores.

_____ 8. Skin temperature may fluctuate between 68° and 104° F without damage.

_____ 9. Rectal temperatures are the same as oral temperatures.

_____ 10. Heat moves down a thermal gradient from a warmer to a cooler region.

_____ 11. Conduction is the transfer of heat using air currents.

_____ 12. The hypothalamus serves as the body's thermostat.

_____ 13. Sweating is an active evaporative heat-loss process.

_____ 14. Most of the blood flow through the skin is for purposes of temperature regulation.

## Fill in the Blank (Answers on page A-30)

15. _____ sympathetic activity to the skin vessels produces heat-conserving vasoconstriction in response to cold exposure.

16. Nonshivering thermogenesis is mediated by the hormones _____ and _____.

17. In response to microbial invasion, certain white blood cells release a chemical known as _____.

18. In response to heat exposure the _____ part of the hypothalamus reduces heat production by _____ skeletal-muscle activity and promotes increased heat loss by inducing skin _____.

19. Aspirin reduces a fever by inhibiting the synthesis of _____.

20. The most common cause of hyperthermia is _____.

21. Sweating is under _____ nervous control.

22. _____ refers to the transfer of heat energy by air currents.

23. The prominent feature in anorexia nervosa is a _____.

24. According to the _____ theory, increased fat storage in adipose tissue signals satiety.

25. Energy output or expenditure falls into two categories: _____ and _____.

26. _____ energy balance exists if the amount of energy in food intake is greater than the amount of energy expended.

27. Chronic diseases such as _____, _____ and _____ are commonly accompanied by a lack of appetite.

28. _____ is the emission of heat energy from the surface of a warm body.

**Matching:** Match the situation or condition to the approximate temperature in ºF.
(Answers on page A-30)

_____ 29. Upper limit compatible with life       a. 100ºF

_____ 30. Central core temperature       b. 68º to 104ºF

_____ 31. Oral temperature       c. 99.7ºF

_____ 32. Skin temperature       d. 110ºF

_____ 33. Rectal temperature       e. 106ºF

_____ 34. Convulsions at this temperature and higher       f. 98.6ºF

**Multiple Choice** (Answers on page A-30)

_____ 35. Which of the following are released in response to endogenous pyrogens?
- a. cachetin
- b. cholecystokinin
- c. prostaglandins
- d. brown fat
- e. electromagnetic waves

_____ 36. Which of the following nervous activities is utilized to control the skin vasomotor responses when the body is subjected to heat exposure?
- a. decreased parasympathetic activity
- b. increased sympathetic activity
- c. increased parasympathetic activity
- d. increased thermoreceptor activity
- e. decreased sympathetic activity

_____ 37. Which of the following nervous activities is utilized to control the skin vasomotor responses when the body is subjected to cold exposure?
- a. decreased parasympathetic activity
- b. increased sympathetic activity
- c. increased parasympathetic activity
- d. increased thermoreceptor activity
- e. decreased sympathetic activity

_____ 38. Which part of the brain houses the body's thermostat?
- a. cerebral cortex
- b. cerebellum
- c. pons
- d. hypothalamus
- e. medulla

_____ 39. Which part of the brain houses the satiety centers?
   a. pons
   b. thalamus
   c. hypothalamus
   d. medulla
   e. pars nervosa

_____ 40. Which of the following activates shivering in response to cold exposure?
   a. pons
   b. thalamus
   c. hypothalamus
   d. medulla
   e. cerebellum

_____ 41. Which of the following is mediated by epinephrine?
   a. shivering
   b. nonshivering thermogenesis
   c. increased glucose utilization
   d. decreased thyroid hormone production
   e. decreased in physical exercise

_____ 42. Which of the following increase carbohydrate utilization?
   a. cachetin
   b. serotonin
   c. dopamine
   d. insulin
   e. inulin

_____ 43. Which of the following contribute to development of emaciation?
   a. glucagon
   b. neuropeptide
   c. cachetin
   d. epinephrine
   e. cholecystokinin

_____ 44. Which of the following is characterized by a low body temperature?
   a. hard exercise
   b. microbial invasion
   c. obesity
   d. anorexia nervosa
   e. cancer

**Point to Ponder**

1. Using common sense and simple logic how do you know that the hypothalamus raises the thermostat setting during fever?
2. Should you take aspirin at the onset of fever?  Why do you have fever in response to microbial invasions?
3. What would you do if your roommate or member of your family developed anorexia nervosa?  Could you help their situation?
4. Should we begin to treat obesity as a disorder or continue with the present social attitude?

**Experiment of the Day**

1. Using your automobile as an "experimental animal" list as many analogies as you can concerning energy balance and thermoregulation.

# ENDOCRINE SYSTEM 15

## CHAPTER OVERVIEW

The two major control systems within the body are the nervous system and the endocrine system. The nervous system deals with rapid responses while the endocrine system provides prolonged sustained control. The nervous system controls the endocrine system primarily through the hypothalamus. The hypothalamus secretes tropic hormones which act on endocrine glands and nontropic hormones that act on nonendocrine target cells. The hypothalamus and endocrine glands secrete hormones or chemical messengers which evoke responses in target cells elsewhere in the body. Synthesis, storage, secretion, transportation, and mechanism of action all depend on the chemical properties of the hormones. There are chemically three types of hormones. Hormone secretion is controlled by negative feedback. Hormones, regardless of their source, utilize the blood to reach their targets.

Hypothalamic neurons secrete two nontropic hormones into capillaries in the posterior pituitary. The hypothalamus also secretes seven tropic hormones which are transported to their target, the anterior pituitary, via the hypothalamo-hypophyseal portal system. Growth, which is under endocrine control, is a typical example of how the nervous system ultimately controls prolonged events. Growth also demonstrates the interaction of hormones which brings about attainment of genetic potential. Homeostasis is a very complex phenomenon.

The role of each body system is equally important in maintaining homeostasis. The peripheral endocrine glands provide an efficient control of nutrient metabolism and mineral balance. The glands in this part of the endocrine system include the thyroid, adrenal, endocrine, pancreas and parathyroid. These glands respond indirectly to the central nervous system and through their various hormones control the metabolic rate, fuel metabolism, and calcium and phosphate balances.  In most instances there are dual controls for increased and decreased activity.

Maintaining adequate nutrient levels to meet the cellular requirements of the organism involves a complex set of hormonal controls.  The peripheral endocrine glands not only meet the needs but respond to excesses and shortages by stimulating the various body systems involved with the storage, release, conversion, and excretion of nutrients and minerals.  Homeostasis is maintained and the cells survive.

# CHAPTER OUTLINE

## Introduction (Text page 479)

The endocrine system is composed of endocrine glands that are scattered throughout the body.

They all accomplish their functions by secreting hormones.

Endocrinology is the study of homeostatic chemical adjustments and other activities accomplished by hormones.

## Comparison of Nervous and Endocrine Systems (Text page 479)

The nervous system is a "wired" system, and the endocrine system is a "wireless" system, even though both systems influence their target cells by means of chemical messengers (page 479).

The endocrine system is one of the body's two major control systems, the other being the nervous system.

In the nervous system, each nerve cell terminates directly as its specific target cells.

The endocrine glands are not anatomically linked with their target cells.

The endocrine chemical messengers are secreted into the blood and delivered to distant target sites.

Specificity of neural communication depends on nerve cells and their target cells having a close anatomical relationship.

Because hormones travel in the blood, they are able to reach virtually all tissues.

Specificity of hormonal action depends on specialization of target cell receptors.

Even though the nervous and endocrine systems have their own realms of authority they are functionally interconnected (page 481).

The nervous system is responsible for coordinating rapid, precise responses.

The endocrine system primarily controls activities that require duration rather than speed including the following: (1) regulatory organic metabolism and water and electrolyte balance, (2) inducing adaptive changes to help the body cope with stressful situations, (3) promoting smooth, sequential growth and development, (4) controlling reproduction, (5) regulating red cell production, and (6) along with the autonomic nervous system, controlling and integrating both circulation and the digestion and absorption of food.

Unlike neurotransmitters, hormones are not quickly removed from the blood.

A hormone may persist in the blood for a few minutes to a matter of hours or even up to a week.

Endocrine effects usually last for some time after the hormone's withdrawal.

The nervous system directly or indirectly controls the secretion of many hormones.

# General Principals of Endocrinology (Text page 484)

Hormones are chemically classified into three categories: peptides, amines, and steroids (page 484).

Hormones are not all similar chemically, but instead fall into three distinct classes according to their biochemical structure: (1) peptides, (2) amines, and (3) steroids.

The means by which a hormone is synthesized, stored, and secreted; the way it is transported in the blood; and the mechanism by which it exerts its effects at the target cell all depend on its chemical properties, the most notable being its solubility.

The following differences in the solubility of the various types of hormones are critical to their function: (1) all peptides and catecholamines are hydrophilic and lipophobic, and (2) all steroid and thyroid hormones are lipophilic and hydrophobic.

The mechanisms of hormone synthesis, storage, and secretion vary according to the class of hormone (page 485).

Peptide hormones are synthesized by the same method used for the manufacture of any protein that is to be exported.

Precursor are synthesized by ribosomes on the rough endoplasmic reticulum.

They migrate to the Golgi complex in membrane-bounded vesicles.

The Golgi complex concentrates the finished hormones, then packages them into secretory vesicles that are pinched off and stored in the cytoplasm until an appropriate signal triggers their secretion.

Upon appropriate stimulation, the secretory vesicles release their contents to the outside by the process of exocytosis.

The following steps are performed by all steroidogenic cells to produce and release their hormonal product.

Cholesterol is the common precursor for all steroid hormones.

Synthesis of the various steroid hormones from cholesterol requires a series of enzymatic reactions that modify the basic cholesterol molecule.

Once formed, the lipid-soluble steroid hormones immediately diffuse through the steroidogenic cell's plasma membrane to enter the blood.

The amines have unique synthetic and secretory pathways.

The amines are derived from the naturally occurring amino acid tyrosine.

Water-soluble hormones are transported dissolved in the plasma, whereas lipid-soluble hormones are largely transported bound to plasma proteins (page 485).

The hydrophilic peptide hormones are transported simply dissolved in the plasma.

The majority of the lipophilic hormones circulate in the blood to their target cell reversibly bound to plasma proteins.

Hormones generally produce their effect by altering intracellular protein activity (page 486).

The hydrophilic peptides and catecholamines bind with specific receptors located on the outer plasma membrane surface of the target cell.

The lipophilic steroids and thyroid hormone easily pass through the surface membrane to bind with receptors located inside the cell.

Hormones exert an effect on their target cell's proteins through three general means:

(1) A few hydrophilic hormones, upon binding with a target cell's surface receptors,

bring about changes in the cell's permeability by altering the shape of adjacent channel-forming proteins already present in the membrane. (2) Most surface-binding hydrophilic hormones function by activating second-messenger systems within the target cells. (3) All lipophilic hormones function by activating specific genes in the target cell to cause the formation of new intracellular proteins, which in turn produce the desired effect.

Because second-messenger systems in general, and cyclic AMP in particular, play such a central role in hydrophilic hormone activity, it is worthwhile to review the steps: (1) Binding of the extracellular first messenger, the hydrophilic hormone, to its surface membrane receptor activates the membrane-bound enzyme adenylate cyclase, which is located on the cytoplasmic side of the plasma membrane. (2) Activated adenylate cyclase converts intracellular ATP to cyclic AMP, the intracellular second messenger. (3) Cyclic AMP triggers a preprogrammed series of biochemical steps that bring about a change in the shape and function of specific preexisting enzymatic proteins in the cell. (4) The altered enzymatic proteins are responsible for bringing about a change in cell activity. The resultant change is the target cell's ultimate physiological response to the hormone. All lipophilic hormones produce their effects in their target cells by enhancing the synthesis of new enzymatic or structural proteins.

Their effects results from stimulation of the cell's genes as follows: (1) Free lipophilic hormone diffuses through the plasma membrane of the target cell and binds with its specific receptor within the nucleus. (2) Once the hormone is bound to the receptor, the hormone-receptor complex binds with DNA at a specific attachment site on the DNA known as the hormone response element. (3) Binding of the hormone-receptor complex with DNA ultimately "turns on" specific genes within the target cell. (4) The activated genes direct the synthesis of new cell protein by producing complementary messenger RNA. (5) The new protein produces the target cell's ultimate physiological response to the hormone.

## The effective plasma concentration of a hormone is normally regulated by changes in its rate of secretion (page 487).

The plasma concentration of a hormone is regulated by adjustments in the rate of its secretion.

Negative feedback is a prominent feature of hormonal control systems.

Many endocrine control systems involve neuroendocrine reflexes, which include neural as well as hormonal components.

The purpose of such reflexes is to produce a sudden increase in hormone secretion in response to a specific stimulus, frequently a stimulus external to the body.

The secretion rates of all hormones rhythmically fluctuate up and down as a function of time.

The most common endocrine rhythm is the diurnal or circadian rhythm, which is characterized by repetitive oscillations in hormone levels that are very regular and have a frequency of one cycle every twenty-four hours.

Endocrine rhythms are locked on or "entrained" to external cues such as the light-dark cycle or the activity cycle.

Endocrine disorders are attributable to hormonal excess, hormonal deficiency, or decreased responsiveness of the target tissue (page 489).

> Endocrine disorders most commonly result from abnormal plasma concentrations of a hormone due to inappropriate rates of secretion; that is, hyposecretion or hypersecretion.

The responsiveness of a target cell to its hormone can be varied by regulating the number of its hormone-specific receptors (page 489).

> A target cell's response to a hormone is correlated with the number of the cell's receptors occupied by molecules of that hormone.
>
> A hormone can influence the activity of another hormone at a given target cell in one of three ways: permissiveness, synergism, and antagonism.

## Hypothalamus and Pituitary (Text page 490)

The pituitary gland consists of anterior and posterior lobes (page 490).

> The pituitary gland or hypophysis, is a small endocrine gland located in a bony cavity at the base of the brain just below the hypothalamus.
>
> The pituitary has two anatomically and functionally distinct lobes, the posterior pituitary and the anterior pituitary.
>
> The posterior pituitary is also termed the neurohypophysis.
>
> The anterior pituitary is also known as the adenohypophysis.

The hypothalamus and posterior pituitary form a neurosecretory system that secretes vasopressin and oxytocin (page 490).

> The hypothalamus and posterior pituitary form a neuroendocrine system that consists of a population of neurosecretory neurons whose cell bodies lie in two well-defined clusters in the hypothalamus and whose axons pass down through the connecting stalk to terminate on capillaries in the posterior pituitary.
>
> The posterior pituitary does not actually produce any hormones.
>
> It simply stores and releases into the blood two small peptide hormones, vasopressin and oxytocin, which are synthesized by the neuronal cell bodies in the hypothalamus.
>
> Vasopressin or antidiuretic hormone, has two major effects that correspond to its two names: (1) it enhances the retention of water by the kidneys, and (2) it causes contraction of arteriolar smooth muscle.
>
> Oxytocin stimulates contraction of the uterine smooth muscle to aid in expulsion of the baby during childbirth, and it promotes ejection of milk from the mammary glands during breast-feeding.

The anterior pituitary secretes six established hormones, many of which are tropic to other endocrine glands (page 491).

> Unlike the posterior pituitary, the anterior pituitary itself synthesizes the hormones that it releases into the blood.
>
> Different cell populations within the anterior pituitary produce and secrete six established peptide hormones: (1) growth hormone, (2) thyroid-stimulating hormone, (3) adrenocorticotropic hormone, (4) follicle-stimulating hormone, (5) luteinizing hormone, and (6) prolactin.

Hypothalamic releasing and inhibiting hormones are delivered to the anterior pituitary by the hypothalamic-hypophyseal portal system to control anterior pituitary hormone secretion (page 493).

> The two most important factors that regulate anterior pituitary hormone secretion are (1) hypothalamic hormones and (2) feedback from target gland hormones.
>
> The secretion of each of the anterior pituitary hormones is stimulated or inhibited by one or more of the seven generally accepted hypophysiotropic hormones.
>
> Depending on their actions, these small peptides are called releasing hormones or inhibiting hormones.
>
> The anatomical and functional link between the hypothalamus and the anterior pituitary is an unusual capillary-to-capillary connection, the hypothalamic-hypophyseal portal system.

In general, feedback by target gland hormones strives to maintain relatively constant rates of anterior pituitary hormone secretion (page 495).

> In most cases, hypophysiotropic hormones initiate a three-hormone sequence: (1) hypothalamic releasing hormone, (2) anterior pituitary tropic hormone, and (3) target endocrine gland hormone.
>
> The target gland hormone also acts to suppress secretion of the tropic hormone that is driving it.

## Endocrine Control of Growth (Text page 495)

Growth depends on growth hormone but is influenced by other factors as well (page 495).

> Growth requires net synthesis of proteins and includes lengthening of long bones as well as increases in the size and number of cells in the soft tissues throughout the body.
>
> An individual's maximum growth capacity is genetically determined.
>
> Attainment of this full growth potential further depends on: (1) an adequate diet, (2) freedom from chronic disease and stressful environmental conditions, and (3) a normal milieu of growth-influencing hormones.

Growth hormone is essential for growth, but it also exerts metabolic effects not related to growth (page 496).

> The overall metabolic effect of growth hormone is to mobilize fat stores as a major energy source while conserving glucose for glucose-dependent tissues such as the brain.
>
> Growth of soft tissues is accomplished by (1) increasing the number of cells (hyperplasia) and (2) increasing the size of cells (hypertrophy).
>
> Growth of the long bones resulting in increased height is the most dramatic effect of growth hormone.
>
> Bones grow in length as a result of proliferation of the cartilage cells in the epiphyseal plates.
>
> Growth in thickness of bone is achieved by the addition of new bone on top of the already existing bone on the outer surface.

Growth hormone exerts its growth-promoting effects indirectly by stimulating somatomedins (page 497).

> The growth-producing actions of growth hormone are not accomplished directly by its effects on the target cells.
>
> These effects are directly brought about by peptide mediators known as somatomedins, whose synthesis in turn is induced by growth hormone.
>
> The major site of somatomedin production is the liver.

Growth hormone secretion is regulated by two hypophysiotropic hormones (page 497).

> Two antagonistic regulatory hormones from the hypothalamus are involved in the control of growth hormone secretion:  growth hormone-releasing hormone and growth hormone-inhibiting hormone.
>
> Growth hormone secretion displays a well-characterized diurnal rhythm.

Abnormal growth hormone secretion results in aberrant growth patterns (page 497).

> Growth hormone deficiency may be caused by a pituitary defect or occur secondarily to hypothalamic dysfunctions.
>
> Hyposecretion of growth hormone in a child results in dwarfism.
>
> Hypersecretion of growth hormone is most often caused by a tumor of the growth hormone-producing cells of the anterior pituitary.
>
> If overproduction of growth hormone begins in childhood the condition is known as gigantism.
>
> If growth hormone hypersecretion occurs after adolescence a disproportionate growth pattern produces a disfiguring condition known as acromegaly.

## Thyroid Gland (Text page 500)

The major thyroid hormone secretory cells are organized into colloid-filled spheres (page 500).

> The chief constituent of the colloid is a large, complex molecule known as thyroglobulin, within which are incorporated the thyroid hormones in their various stages of synthesis.
>
> The follicular cells produce two iodine-containing hormones derived from the amino acid tyrosine:  tetraiodothryronine (thyroxine) and triiodothyronine.
>
> These two hormones, collectively referred to as thyroid hormone, are important regulators of overall basal metabolic rate.
>
> Interspersed in the interstitial spaces between the follicles is another secretory type cell, the C cells, so called because they secrete the peptide hormone calcitonin, which plays a role in calcium metabolism.

All of the steps of thyroid hormone synthesis occur on the large thyroglobulin molecule, which subsequently stores the hormones (page 501).

> The synthesis, storage, and secretion of thyroid hormone involve the following steps:
> (1) Tyrosine becomes incorporated in the much larger thyroglobulin molecules which are exported from the follicular cells into the colloid by exocytosis.  (2) The thyroid captures iodine from the blood and transfers it into the colloid by means of a very active "iodine pump."  (3) Within the colloid iodine is quickly attached to a tyrosine within the thyro-

globulin molecule.
Attachment of one iodine to tyrosine yields monoiodotyrosine (MIT).
Attachment of two iodines to tyrosine yields diiodotyrosine (DIT). (4) Coupling of two DITs yields tetraiodothyronine (thyroxine).
Coupling of one MIT and one DIT yields triiodothyronine.

## The follicular cells phagocytize thyroglobulin-laden colloid to accomplish thyroid hormone secretion (page 502).

Upon appropriate stimulus for thyroid hormone secretion, the follicular cells internalize a portion of the thyroglobulin-hormone complex by phagocytizing a piece of colloid. The biologically active thyroid hormones are split off.
The MIT and DIT are of no endocrine value.

## Most of the secreted $T_4$ is converted into $T_3$ outside the thyroid (page 502).

Triiodothyronine is the major biologically active form of thyroid hormone at the cellular level, even though the thyroid secretes mostly tetraiodothyronine.

## Thyroid hormone is the primary determinant of the body's overall metabolic rate and is also important for bodily growth and normal development and function of the nervous system (page 502).

Virtually every tissue in the body is affected either directly or indirectly by thyroid hormone.
Thyroid hormone increases the body's overall basal metabolic rate.
Closely related to thyroid hormone's overall metabolic effect is its calorigenic effect.
Thyroid hormone increases target-cell responsiveness to catecholamines, the chemical messengers used by the sympathetic nervous system and its hormonal reinforcements from the adrenal medulla.
Thyroid hormone increases heart rate and force of contraction, thus increasing cardiac output.
Thyroid hormone not only stimulates growth hormone secretion but also promotes the effects of growth hormone on the synthesis of new structural proteins and on skeletal growth.
Thyroid hormone plays a crucial role in the normal development of the nervous system.

## Thyroid hormone is regulated by the hypothalamus-pituitary-thyroid axis (page 502).

Thyroid-stimulating hormone, the thyroid tropic hormone from the anterior pituitary, is the most important physiological regulator of thyroid hormone secretion.
Thyroid hormone, in negative-feedback fashion, "turns off" TSH secretion, whereas the hypothalamic thyrotropin-releasing hormone (TRH), in tropic fashion "turns on" TSH secretion by the anterior pituitary.

## Abnormalities of thyroid function include both hypothyroidism and hyperthyroidism (page 503).

Hypothyroidism can result (1) from primary failure of the thyroid gland itself; (2) secondary to a deficiency of TRH, TSH, or both; or (3) from an inadequate dietary supply of iodine.

The most common cause of hyperthyroidism is Grave's disease, an autoimmune disease in which the body erroneously produces thyroid-stimulating immunoglobulin (TSI), an antibody whose target is the TSH receptors on the thyroid cells.

Less frequently, hyperthyroidism occurs secondary to excess TRH or TSH or in association with a hypersecreting thyroid tumor.

## A goiter may or may not accompany either hypothyroidism or hyperthyroidism (page 504).

A goiter refers to an enlarged thyroid gland.

Hypothyroidism secondary to hypothalamic or anterior pituitary failure will not be accompanied by a goiter because the thyroid is not being adequately stimulated.

With hypothyroidism caused by thyroid gland failure or lack of iodine, a goiter does develop because the circulating level of thyroid hormone is so low that there is little negative-feedback inhibition on the anterior pituitary, and TSH secretion is therefore elevated.

Excessive TSH secretion resulting from hypothalamic or anterior pituitary defect would obviously be accompanied by a goiter because of overstimulation of thyroid growth.

In Grave's disease, a hypersecreting goiter occurs because TSI promotes growth of the thyroid as well as enhances secretion of thyroid hormone.

Hyperthyroidism resulting from overactivity of the thyroid in the absence of over-stimulation, such as that caused by an uncontrolled thyroid tumor, is not accompanied by a goiter.

# Adrenal Glands (Text page 505)

## Each adrenal gland consists of an outer, steroid-secreting adrenal cortex and an inner, catecholamine-secreting adrenal medulla (page 505).

Each adrenal is actually composed of two endocrine organs, the adrenal cortex surrounding the adrenal medulla.

## The adrenal cortex secretes mineralocorticoids, glucocorticoids, and sex hormones (page 505).

The adrenal cortex produces a number of different adrenocortical hormones, all of which are steroids derived from the common precursor molecule, cholesterol.

Mineralocorticoids influence mineral balance.

Glucocorticoids, primarily cortisol, play a major role in glucose metabolism as well as in protein and lipid metabolism.

The sex hormones are identical or similar to those produced by the gonads.

## Mineralocorticoids' major effects are on electrolyte balance and blood pressure homeostasis (page 505).

Mineralocorticoids are essential for life.

Without aldosterone, a person rapidly dies from circulatory shock because of a marked fall in plasma volume.

Aldosterone secretion is increased by (1) activation of the renin-angiotensin-aldosterone system by factors related to a reduction in sodium and a fall in blood pressure and (2) direct stimulation of the adrenal cortex by a rise in plasma potassium concentration.

## Glucocorticoids exert metabolic effects and have an important role in adaptation to stress (page 506).

Cortisol stimulates hepatic gluconeogenesis.

Cortisol inhibits glucose uptake and use by many tissues, but not the brain, thus sparing glucose for use by the brain, which absolutely requires it as metabolic fuel.

Cortisol stimulates protein degradation in many tissues, especially muscle.

Cortisol facilitates lipolysis, the breakdown of lipid stores in adipose tissue, thus releasing free fatty acids into the blood.

Cortisol must be present in adequate amounts to permit catecholamines to induce vasoconstriction.

Cortisol plays a key role in adaptation to stress.

Administration of large amounts of glucocorticoids inhibits almost every step of the inflammatory response.

## Cortisol secretion is directly regulated by ACTH (page 507).

Cortisol secretion by the adrenal cortex is regulated by a negative-feedback system involving the hypothalamus and anterior pituitary.

## The adrenal cortex secretes both male and female sex hormones in both sexes (page 508).

In both sexes, the adrenal cortex produces both androgens.

The only adrenal sex hormone that has any biological importance is the androgen dehydroepiandrosterone.

This adrenal androgen is responsible for androgen-dependent processes in the female.

## The adrenal gland may secrete too much or too little of any one of its hormones (page 508).

Excess mineralocorticoid secretion may be caused by (1) a hypersecreting adrenal tumor made up of aldosterone-secreting cells or (2) inappropriately high activity of the renin-angiotensin system.

Excessive cortisol secretion can be caused by (1) overstimulation of the adrenal cortex by excessive amounts of CRH and/or ACTH; (2) adrenal tumors that uncontrollably secrete cortisol independent of ACTH.

Excess adrenal androgen secretion, a masculinizing condition, is more common than the extremely rare feminizing condition of excess adrenal estrogen secretion.

The adrenogenital syndrome is most commonly caused by an inherited enzymatic defect in the cortisol steroidogenic pathway.

In primary adrenocortical insufficiency, also known as Addison's disease, all layers of the adrenal cortex are undersecreting.

Secondary adrenocortical insufficiency may occur because of a pituitary or hypothalamic abnormality, resulting in insufficient ACTH secretion.

## The catecholamine-secreting adrenal medulla is a modified sympathetic postganglionic neuron (page 509).

The adrenal medulla is actually a modified part of the sympathetic nervous system.

The adrenal medulla is composed of modified postganglionic sympathetic neurons.

Postganglionic sympathetic neurons in the adrenal medulla do not possess axonal fibers that terminate on effector organs.

The ganglionic cell bodies within the adrenal medulla release their chemical transmitter directly into the circulation upon stimulation by the preganglionic fiber.

## Epinephrine reinforces the sympathetic nervous system and exerts additional metabolic effects as well (page 509).

The sympathetic nervous system and adrenomedullary epinephrine mobilize the body's resources to support peak physical exertion.

Epinephrine, circulating in the blood, is able to reach target cells that are not innervated by the sympathetic nervous system.

Epinephrine prompts the mobilization of stored carbohydrate and fat to provide immediately available energy for use as needed to fuel muscular work.

## Adrenomedullary dysfunction is very rare (page 510).

No adverse effects have been attributed to a deficiency of epinephrine.

The only catecholamine disorder is pheochromocytoma, a rarely occurring catecholamine secreting tumor.

## The stress response is a generalized, nonspecific pattern of neural and hormonal reactions to any situation that threatens homeostasis (page 510).

Besides epinephrine, a number of other hormones are involved in the overall stress response.

The predominant hormonal response is activation of the CRH-ACTH cortisol system.

Cortisol breaks down fat and protein stores while expanding carbohydrate stores.

Vasopressin secretion is also increased during stressful situations.

## The multifaceted stress response is coordinated by the hypothalamus (page 511).

The hypothalamus receives input concerning physical and emotional stressors from virtually all areas of the body and from many receptors throughout the body.

In response, the hypothalamus directly activates the sympathetic nervous system, secretes CRH to stimulate ACTH and cortisol release, and triggers the release of vasopressin.

Sympathetic stimulation brings about the secretion of epinephrine.

## Activation of the stress response by chronic psychosocial stressors may be harmful (page 511).

There is strong circumstantial evidence for a link between chronic exposure to psycho-social stressors and the development of pathological conditions such as atherosclerosis and high blood pressure, although no definitive cause-and-effect relationship has been ascertained.

# Endocrine Control of Fuel Metabolism (Text page 512)

## All three classes of nutrient molecules can be used to provide cellular energy and, to a large extent, can be interconverted (page 512).

Anabolism refers to the buildup or synthesis of larger organic macromolecules from the small organic molecular subunits.

Catabolism refers to the breakdown or degradation of large, energy-rich organic molecules within cells.

Mobilized glucose, fatty acid, and amino acid molecules can be used as needed for energy production or cellular synthesis elsewhere in the body.

Essential nutrients such as the essential amino acids and vitamins cannot be formed in the body by conversion from another type of organic molecule.

## The brain must be continuously supplied with glucose, even between meals when no new nutrients are being taken up from the digestive tract (page 514).

Excess circulating glucose is stored as glycogen.

Additional glucose is transformed into fatty acids and glycerol, which are used to synthesize triglycerides.

Excess circulating fatty acids derived from dietary intake also become incorporated into triglycerides.

Excess circulating amino acids not needed for protein synthesis are converted to glucose and fatty acids.

The major site of energy storage for excess nutrients of all three classes is adipose tissue.

A substantial amount of energy is stored as structural protein, primarily in muscle, the most abundant protein mass in the body.

The brain normally depends on the delivery of adequate blood glucose as its sole source of energy.

When new dietary glucose is not entering the blood, tissues not obligated to use glucose shift their metabolic gears to burn fatty acids instead, thus sparing glucose for the brain.

## Metabolic fuels are stored during the absorptive state and are mobilized during the postabsorptive state (page 515).

Ingested nutrients are being absorbed and entering the blood during the absorptive state or fed state.

During this time, glucose is plentiful and serves as the major energy source.

The average meal is completely absorbed in about four hours.

On a typical three-meals-a-day diet, no nutrients are being absorbed from the digestive tract during late morning, late afternoon, and throughout the night.

These times constitute the postabsorptive state, or fasting state.

During the absorptive state, the glut of absorbed nutrients is swiftly removed from the blood and placed into storage; during the postabsorptive state, these stores are catabolized to maintain the blood concentrations at levels necessary to sustain tissue energy demands.

## The pancreatic hormones, insulin and glucagon, are the most important in regulating fuel metabolism (page 515).

Scattered throughout the pancreas are clusters of endocrine cells known as the islets of Langerhans.

The most abundant pancreatic endocrine-cell type is the beta cell, the site of insulin synthesis and secretion.

Next most important are the alpha cells, which produce glucagon.

Insulin lowers blood glucose, amino acid, and fatty acid levels and promotes anabolism of these small nutrient molecules (page 515).

Insulin lowers the blood levels of glucose, fatty acids, and amino acids and promotes their storage.

Insulin facilitates glucose transport into most cells.

Insulin stimulates glycogenesis and inhibits glycogenolysis.

Insulin further decreases hepatic glucose output by inhibiting gluconeogenesis, the conversion of amino acids into glucose in the liver.

Insulin increases the transport of glucose into adipose-tissue cells where glucose serves as a precursor for the formation of fatty acids and glycerol.

Insulin activates enzymes that catalyze the production of fatty acids from glucose derivatives.

Insulin promotes the entry of fatty acids from the blood into adipose-tissue cells.

Insulin inhibits lipolysis, thus reducing the release of fatty acids from adipose tissue into the blood.

Insulin promotes the active transport of amino acids from the blood into muscles and other tissues.

Insulin increases the rate of amino acid incorporation into protein by stimulating the cell's protein synthesizing machinery.

Insulin inhibits protein degradation.

The primary stimulus for increased insulin secretion is an increase in blood glucose concentration (page 516).

The primary control of insulin secretion is a direct negative-feedback system between the pancreatic beta cells and the concentration of glucose in the blood flowing to them.

An elevated plasma amino acid level directly stimulates the beta cells to increase insulin secretion.

The major gastrointestinal hormones, especially gastric inhibitory peptide, stimulate pancreatic insulin secretion.

The increase in parasympathetic activity that occurs in response to food in the digestive tract stimulates insulin release.

Sympathetic stimulation and the concurrent increase in epinephrine both inhibit insulin secretion.

The are two types of diabetes mellitus, depending on the insulin-secreting capacity of the cells (page 517).

Diabetes mellitus is by far the most common of all endocrine disorders.

The most prominent feature of diabetes mellitus is hyperglycemia.

A large urine volume occurs both in diabetes mellitus due to insulin insufficiency and in diabetes insipidus due to vasopressin deficiency.

Type I diabetes is characterized by a lack of insulin secretion.

Type II diabetes is characterized by normal or even increased insulin secretion but reduced sensitivity of insulin's target cells to its presence.

The symptoms of diabetes mellitus are characteristic of an exaggerated postabsorptive state (page 518).

The acute consequences of diabetes mellitus can be grouped according to the effects of inadequate insulin action on carbohydrate, fat, and protein metabolism.

In the fasting state, hyperglycemia arises from reduced glucose uptake by cells.
Glucose appears in the urine producing osmotic diuresis.
Excess fluid loss leads to dehydration which can lead to circulatory failure.
Circulatory failure can lead to death due to low cerebral blood flow or renal failure.
Cells shrink due to dehydration which can lead to nervous system malfunction.
Increased thirst and food intake are characteristic of diabetes mellitus.
Weight loss occurs.
Triglyceride synthesis decreases while lipolysis increases.
The liver releases excessive ketones into the blood which leads to acidosis.
Diabetes mellitus is a complicated disease that can produce repercussions on the circulatory system, kidneys, respiratory system and nervous system.

## Insulin excess causes brain-starving hypoglycemia (page 518).

Insulin excess can occur in a diabetic patient when too much insulin has been injected for the person's caloric intake and exercise level, resulting in so-called insulin shock.
An abnormally high blood insulin level may occur in a nondiabetic individual with a beta cell tumor or in whom the beta cells are overresponsive to glucose.
The consequences of insulin excess are primarily manifestations of the effects of hypoglycemia on the brain.

## Glucagon in general opposes the actions of insulin (page 520).

Many physiologists view the insulin-secreting beta cells and the glucagen-secreting alpha cells as a coupled endocrine system whose combined secretory output is a major factor in the regulation of fuel metabolism.
Glucagon affects many of the same metabolic processes that are influenced by insulin, but in most cases glucagon's actions are opposed to those of insulin.

## Glucagon secretion is increased during the postabsorptive state (page 520).

Glucagon secretion is increased during the postabsorptive state and decreased during the absorptive state.
The major factor regulating glucagon secretion is a direct effect of the blood glucose concentration on the endocrine pancreas.
There is a direct negative-feedback relationship between blood glucose concentration and the alpha cell's rate of secretion.
An elevated blood glucose level inhibits glucagon secretion.
Glucagon and insulin work together homeostatically to restore blood glucose levels to normal.

## Glucagon excess can aggravate the hyperglycemia of diabetes mellitus (page 521).

No known clinical abnormalities are attributable to glucagon deficiency or excess per se.
Since glucagon is a hormone that raises blood glucose, its excess intensifies the hyperglycemia of diabetes mellitus.

## Epinephrine, cortisol, growth hormone, and thyroid hormone also exert direct metabolic effects (page 521).

The stress hormones, epinephrine and cortisol, both increase blood levels of glucose and fatty acids.
Cortisol mobilizes amino acids by promoting protein catabolism.

Growth hormone has protein anabolic effects in muscle.

Glucagen, epinephrine, cortisol, and growth hormone are considered insulin antagonists.

# Endocrine Control of Calcium Metabolism (Text page 522)

Plasma calcium must be closely regulated to prevent changes in neuromuscular excitability (page 522).

Three hormones---parathyroid hormone, calcitonin, and vitamin D---control calcium and phosphate metabolism.

Extracellular fluid calcium plays a vital role in a number of essential activities, the most important of which is its effect on neuromuscular excitability.

In hypocalcemia excitable tissues may be brought to threshold by normally ineffective physiological stimuli.

Hypercalcemia causes cardiac arrhythmias.

Regulation of calcium metabolism depends on hormonal control of exchanges between the ECF and three other compartments:  bone, kidneys, and intestine.

Parathyroid hormone raises free plasma calcium levels by its effects on bone, kidneys, and intestine (page 522).

Parathyroid hormone (PTH) is a peptide hormone secreted by the parathyroid glands.

The overall effect of PTH is to increase the calcium concentration of plasma.

Parathyroid hormone also acts to lower plasma phosphate concentration.

Parathyroid hormone uses bone as a "bank" from which it withdraws calcium as needed to maintain the plasma calcium level.

Parathyroid hormone stimulates calcium conservation and promotes phosphate elimination by the kidneys during the formation of urine.

The primary regulator of PTH secretion is the plasma concentration of free calcium (page 523).

Parathyroid hormone secretion is increased in response to a fall in plasma calcium concentration and decreased by a rise in plasma calcium levels.

Calcitonin lowers the plasma calcium concentration but is not important in the normal control of calcium metabolism (page 524).

Calcitonin, the hormone produced by the C cells of the thyroid gland, also exerts an influence on plasma calcium levels.

Since calcitonin reduces plasma calcium levels, this system constitutes a second simple negative-feedback control over plasma calcium concentration, one that is opposed to the PTH system.

Vitamin D is actually a hormone that increases calcium absorption in the intestine (page 524).

The final factor involved in the regulation of calcium metabolism is cholecalciferol, or vitamin D, a steroidlike compound that's essential for calcium absorption in the intestine.

The skin is actually an endocrine gland and vitamin D a hormone.

The most dramatic and biologically important effect of activated vitamin D is to increase calcium absorption in the intestine.

Disorders in calcium metabolism may arise from abnormal levels of parathyroid hormone or vitamin D (page 524).

Hypercalcemia reduces the excitability of muscle and nervous tissue.

Excessive mobilization of calcium and phosphate from skeletal stores leads to thinning of bone.

An increased incidence of formation of calcium-containing kidney stones occurs because of the excess quantity of calcium being filtered through the kidneys.

Hypercalcemia can cause digestive disorders such as peptic ulcers, nausea, and constipation.

Hypoparathyroidism leads to hypocalcemia and hyperphosphatemia.

The major consequence associated with vitamin D deficiency is impaired intestinal absorption of calcium.

The rates of bone formation and reabsorption are about equal throughout most of adult life.

A reduction in bone mass is known as osteoporosis.

The underlying cause of osteoporosis is uncertain.

## KEY TERMS

Absorptive state-p515
Addison's disease-p509
Adenohypophysis-p490
Adrenal cortex-p505
Adrenal medulla-p505
Adrenocorticotropic hormone-p491
Amine hormones-p484
Antagonism-p489
Calcitonin-p501
Circadian rhythm-p489
Conn's syndrome-p508
Cushings syndrome-p508
Diurnal rhythm-p489
Follicle-stimulating hormone-p491
Glucagon-p520
Glucocorticoids-p505
Gluconeogenesis-p516
Glycogenesis-p516

Glycogenolysis-p516
Goiter-p504
Growth hormone-p491
Hormone response element-p487
Hypercalcemia-p522
Hyperglycemia-p517
Hyperplasia-p496
Hypersecretion-p489
Hyperthyroidism-p504
Hypertrophy-p496
Hypocalcemia-p522
Hypoglycemia-p518
Hypophysiotropic hormone-p493
Hypophysis-p490
Hyposecretion-p489
Hypothyroidism-p503
Insulin-p515
Luteinizing hormone-p491

Mineralocortcoids-p505
Neuroendocrine reflex-p488
Neurohypophysis-p490
Nontropic hormones-p492
Osteoporosis-p526
Peptide hormones-p484
Permissiveness-p489
Postabsorptive state-p515
Prolactin-p491
Somatostatin-p497
Steroid hormones-p484
Synergism-p489
Tetraiodothyronine-p501
Thyroglobulin-p501
Thyroid-stimulating hormone-p491
Thyroxine-p501
Triiodothyronine-p501
Tropic hormone-p491

## REVIEW EXERCISES

**True - False** (Answers on page A-31)

_____    1. Bone is a living tissue.

_____    2. Bones grow in length as a result of proliferation of the cartilage cells in the epiphyseal plates.

_____    3. The major site of somatomedin production is the liver.

_____ 4. Growth hormone deficiency may be caused by a pituitary defect.

_____ 5. Thyroid hormone is believed to be an important growth-promoting factor.

_____ 6. Androgens ultimately terminate linear growth by stimulating complete conversion of the epiphyseal plates to bone.

_____ 7. The posterior pituitary does not actually produce any hormones.

_____ 8. The posterior pituitary is also known as the adenohypophysis.

_____ 9. Oxytocin causes contraction of arteriolar smooth muscle.

_____ 10. Prolactin enhances breast development and milk production in females.

_____ 11. The most intimate means of intracellular communication is by gap junctions.

_____ 12. A tropic hormone primarily exerts its effects on nonendocrine target tissues.

_____ 13. A single hormone may be secreted by only one endocrine gland.

_____ 14. All steroid and thyroid hormones are hydrophilic.

_____ 15. Cholesterol is the common precursor for all steroid hormones.

_____ 16. The majority of the hydrophilic hormones circulate in the blood to their target cells reversibly bound to plasma proteins.

_____ 17. Negative feedback exists when the output of a system opposes a change in input.

_____ 18. The MIT and DIT are of no endocrine value.

_____ 19. $T_4$ is the major biologically active form of thyroid hormone at the cellular level.

_____ 20. Thyroid hormone increases the body's overall basal metabolic rate.

_____ 21. Thyroid hormone decreases target-cell responsiveness to catecholamines.

_____ 22. If an individual has hypothyroidism from birth, he has a condition known as myxedema.

_____ 23. A hypersecreting goiter occurs because TSI promotes growth of the thyroid as well as enhancing secretion of thyroid hormone.

_____ 24. Adrenal means next to the kidney.

_____ 25. About twenty percent of the adrenal gland is composed of the cortex.

_____ 26. Without aldosterone, a person rapidly dies from circulatory shock because of the marked fall in plasma volume.

_____ 27. Cortisol stimulates protein degradation in many tissues, except muscle.

_____ 28. A true hermaphrodite has the gonads of both sexes.

_____ 29. Addison's disease is also known as secondary adrenocortical insufficiency.

_____ 30. Excess circulating glucose is stored as glycogen.

_____ 31. The average meal is completely absorbed in about ten hours.

_____ 32. The beta cell is the site of insulin synthesis and secretion.

_____ 33. Diabetes mellitus is by far the most common of all endocrine disorders.

_____ 34. Insulin is the only hormone capable of lowering blood glucose levels.

_____ 35. Type II diabetes is characterized by a lack of insulin secretion.

_____ 36. Glucagon secretion is increased during the postabsorptive state and decreased during the absorptive state.

_____ 37. An elevated blood glucose level inhibits glucagon secretion.
_____ 38. The overall effect of PTH is to decrease the calcium concentration of plasma.
_____ 39. Parathyroid hormone uses bone as a "bank."
_____ 40. PTH secretion is increased in response to a fall in plasma calcium concentration and decreased by a rise in plasma calcium levels.
_____ 41. Calcitonin increases plasma calcium levels.
_____ 42. Vitamin D is a hormone.
_____ 43. Phosphate is regulated directly by vitamin D.
_____ 44. A fall in plasma phosphate causes a decrease in plasma calcium.
_____ 45. Hypercalcemia reduces the excitability of muscle and nervous tissue.

**Fill in the Blank** (Answers on page A-32)
46. FSH and LH are collectively referred to as _____.
47. The _____ is a small endocrine gland located in a bony cavity at the base of the brain just below the hypothalamus.
48. _____ contributes to the pubertal growth spurt by promoting protein synthesis and bone growth.
49. The bone cells that produce the organic matrix are known as _____.
50. _____ is similar to bone, except that it is not calcified.
51. A long bone basically consists of a fairly uniform cylindrical shaft, the _____ with a flared articulating knob at either end, an _____.
52. Hyposecretion of growth hormone in a child results in _____.
53. Overproduction of growth hormone beginning in childhood results in
_____.
54. _____ are long-range chemical messengers that are specifically secreted into the blood by endocrine glands in response to an appropriate signal.
55. _____ stimulates growth hormone secretion.
56. _____ regulates skin coloration by controlling the dispersion of granules containing the pigment _____.
57. Growth of soft tissues is accomplished by (1) increasing the number of cells (_____) by stimulating cell division and (2) increasing the size of cells (_____) by favoring synthesis of proteins.
58. The _____, which secretes blood-borne chemical messengers known as hormones, specializes in intracellular communications.
59. Once the hormone is bound to the receptor, the hormone-receptor complex binds with DNA at a specific attachment site on DNA known as the _____.
60. A hormone that has as its primary function the regulation of hormone secretion by another endocrine gland is classified functionally as a _____ hormone.
61. _____ which include the hormones secreted by the adrenal cortex, gonads, and most placental hormones, are neutral lipids derived from cholesterol.

62. The secretion rates of all hormones rhythmically fluctuate up and down as a function of time.  The most common endocrine rhythm is the _____, or _____, which is characterized by repetitive oscillations in hormone levels that are very regular and have a frequency of one cycle every twenty-four hours.

63. With _____, one hormone must be present in adequate amounts for the full exertion of another hormone's effect.

64. _____ occurs when the actions of several hormones are complementary and their combined effect is greater than the sum of their separate effects.

65. _____ occurs when one hormone causes the loss of another hormone's receptors, reducing the effectiveness of the second hormone.

66. The secretion of each of the anterior pituitary hormones is stimulated or inhibited by one or more of the seven generally accepted hypothalamic _____ _____.

67. The anatomical and functional link between the _____ and _____ is an unusual capillary-to-capillary connection called the hypothalamo-hypophyseal portal system.

68. The _____ --- thyroid hormone and adrenomedullary catecholamines --- have unique synthetic and secretory pathways.

69. If growth hormone hypersecretion occurs after adolescence the disproportionate growth pattern produces a disfiguring condition known as _____.

70. _____ refers to all of the chemical reactions that occur within the cells of the body.

71. _____ refers to the build-up or synthesis of larger organic macro-molecules from the small organic molecular subunits.

72. _____ refers to the breakdown or degradation of large, energy-rich organic molecules within cells.

73. Scattered throughout the pancreas are clusters of endocrine cells known as the _____.

74. _____ is the production of glycogen.

75. Hypoglycemia refers to _____ blood glucose.

76. _____, the hormone produced by the C-cells of the thyroid gland, exerts an influence on plasma calcium levels.

77. The final factor involved in the regulation of calcium metabolism is _____.

78. When demineralized bones become soft and deformed, bowing to the pressures of weight bearing, is known as _____ in children and _____ in adults.

79. A reduction in bone mass is known as _____.

80. The chief constituent of the colloid is a large, complex molecule known as _____.

81. _____ is an autoimmune disease in which the body erroneously produces thyroid-stimulating immunoglobulin (TSI).
82. A _____ refers to an enlarged thyroid gland.
83. _____ are so called because they secrete the peptide hormone calcitonin.
84. MIT is short for _____.
85. $T_4$ is short for _____.
86. The three layers of the cortex are; (1) _____ -the outermost layer, (2) _____ -middle and largest layer, (3) _____ _____ -the innermost layer.
87. _____ plays an important role in carbohydrate, protein, and fat metabolism.
88. _____ syndrome is an excessive cortisol secretion.
89. _____ refers to the condition where a woman develops a male pattern of body hair.
90. _____ disease is where all cells of the adrenal cortex are under-secreting.
91. Epinephrine and norepinephrine are stored in _____.
92. The only catecholamine disorder is a _____, a rarely occurring catecholamine-secreting tumor.

**Matching:** Match the endocrine gland to its hormone (Answers on page A-33)

_____ 93. Cortisol
_____ 94. Oxytocin
_____ 95. Vitamin D
_____ 96. Estrogen
_____ 97. Melatonin
_____ 98. Glucagon
_____ 99. Inhibin
_____ 100. Testosterone
_____ 101. Thymosin
_____ 102. Prolactin
_____ 103. TRH
_____ 104. Growth hormone
_____ 105. Calcitonin
_____ 106. Chorionic gonadotropin
_____ 107. Gastrin
_____ 108. Atrial natriuretic peptide
_____ 109. FSH
_____ 110. CRH
_____ 111. Thyroxine
_____ 112. Aldosterone
_____ 113. Progesterone

a. adrenal medulla
b. posterior pituitary
c. islets of Langerhans
d. pineal gland
e. hypothalamus
f. thyroid follicular cells
g. kidneys
h. duodenum
i. heart
j. placenta
k. skin
l. thymus
m. liver
n. anterior pituitary
o. stomach
p. parathyroid
q. thyroid C cells
r. gonads
s. adrenal cortex

_____ 114. PTH
_____ 115. Insulin
_____ 116. Somatomedins
_____ 117. LH
_____ 118. GHIH
_____ 119. Epinephrine
_____ 120. TSH
_____ 121. Renin
_____ 122. ICSH
_____ 123. Androgens
_____ 124. ACTH
_____ 125. Vasopressin
_____ 126. PIH
_____ 127. Norepinephrine
_____ 128. Cholecystokin
_____ 129. Erythropoietin
_____ 130. PRH
_____ 131. GHRH
_____ 132. Secretin

Match the various causes of thyroid dysfunction to the plasma concentrations of hormones and presence of goiters (Answers on page A-33)

_____ 133. Primary failure of thyroid gland
_____ 134. Anterior pituitary failure or hypothalamic failure
_____ 135. Excess hypothalamic secretion or excess pituitary secretion
_____ 136. Lack of dietary iodine
_____ 137. Hypersecreting thyroid tumor
_____ 138. Grave's disease

a. increased thyroid hormone, decreased TSH and no goiter
b. decreased thyroid hormone, increased TSH with goiter
c. increased thyroid hormone, decreased TSH with goiter
d. decreased thyroid hormone, decreased TRH and/or decreased TSH and no goiter
e. increased thyroid hormone, increased TRH and/or increased TSH with goiter

Match the condition to the disorder (Answers on page A-34)

_____ 139. Excess aldosterone
_____ 140. Excess androgen
_____ 141. Deficient cortisol and aldosterone
_____ 142. Excess cortisol
_____ 143. Deficient cortisol

a. Cushing syndrome
b. Addison's disease
c. Conn's syndrome
d. Secondary adrenocortical insufficiency
e. Adrenogenital syndrome

**Multiple Choice** (Answers on page A-35)

_____ 144.  Which of the following are released into the blood specifically by neurosecretory neurons?
   a.  neurotransmitters
   b.  paracrines
   c.  neurohormones
   d.  neuromodulator
   e.  hormones

_____ 145.  Which of the following hormones has as its target an endocrine gland?
   a.  paracrines
   b.  tropic hormones
   c.  nontropic hormones
   d.  neurotransmitters
   e.  neuromodulators

_____ 146.  Which of the following promotes breast development and stimulates milk
secretion?
   a.  Leutinizing hormone
   b.  Somatomedins
   c.  Renin
   d.  Oxytocin
   e.  Prolactin

_____ 147.  Which of the following decreases plasma calcium concentration?
   a.  Parathyroid hormone
   b.  Glucagon
   c.  Cortisol
   d.  Calcitonin
   e.  Oxytocin

_____ 148.  Which of the following is a condition resulting from hypersecretion of growth
hormone after adolescence?
   a.  Laron dwarfism
   b.  achondroplastic dwarfism
   c.  gigantism
   d.  acromegaly
   e.  exophthalmic goiter

_____ 149.  Which of the following hormones increase uterine contractility?
   a.  oxytocin
   b.  vasopressin
   c.  PRH
   d.  GHRH
   e.  none of the above

_____ 150. Which of the following hormones produces vasoconstriction in arterioles?
     a. oxytocin
     b. aldosterone
     c. cortisol
     d. ACTH
     e. vasopressin

_____ 151. Which of the following condition occurs when hypersecretion of growth hormone begins in early childhood?
     a. dwarfism
     b. gigantism
     c. acromegaly
     d. Laron gigantism
     e. achondroplastic gigantism

_____ 152. Which of the following refers to a potassium depletion?
     a. hypocalcemia
     b. osteomalacia
     c. hypokalemia
     d. osteoporosis
     e. pheochromocytoma

_____ 153. Which of the following refers to a reduction in bone mass?
     a. hirsutism
     b. osteoporosis
     c. osteomalacia
     d. hypocalcemia
     e. hypokalemia

_____ 154. Which of the following refers to the prominent bulging eyes characteristic of Grave's disease?
     a. goiter
     b. hirsutism
     c. osteomalacia
     d. optic protrusia
     e. exophthalmos

_____ 155. Which disorder is characterized by a lack of insulin?
     a. Type I diabetes
     b. Type II diabetes
     c. Diabetes insipidus
     d. pheochromocytoma
     e. insulemia

_____ 156. Which of the following is characterized by the development of a male pattern of body hair in women?
    a. pheochromocytoma
    b. osteomalacia
    c. goiter
    d. exophthalmos
    e. hirsutism

_____ 157. Which of the following disorders is characterized by the demineralization of bones in adults?
    a. rickets
    b. osteoporosis
    c. osteomalacia
    d. hirsutism
    e. pheochromocytoma

_____ 158. Which of the following disorders is characterized by a catecholamine-secreting tumor?
    a. pheochromocytoma
    b. catechofibroma
    c. catecholemia
    d. exophthalmos
    e. osteomalacia

_____ 159. Which disorder is characterized by low blood calcium?
    a. hypokalemia
    b. hypercalcemia
    c. hypocalcemia
    d. osteoporosis
    e. osteomalacia

_____ 160. Which disorder refers to an enlarged thyroid gland?
    a. hypothyroidism
    b. hyperthyroidism
    c. pheochromocytoma
    d. exophthalmos
    e. goiter

_____ 161. Which disorder is characterized by reduced sensitivity of insulin's target cells?
    a. insulemia
    b. type I diabetes
    c. type III diabetes
    d. diabetes insipidus
    e. type II diabetes

## Points to Ponder

1. Why do hypothalamic neurons extend into the posterior pituitary but not into the anterior pituitary?
2. Why must there be releasing and inhibiting hormones secreted by the hypothalamus? Why not evolve a system whereby the endocrine gland responds in the negative feedback loop?
3. How does growth fit in homeostasis when homeostasis means a state of physiological equilibrium produced by a balance of functions and chemical composition within an organism.
4. Recent research indicates that diabetes mellitus is genetic. If this is true then the use of insulin only increases the frequency of the disease in the human population. How do we reduce the frequency of the disease and still help those who have the disorder?
5. How much of your behavior can you attribute to the hormones discussed in this chapter?
6. What do you think causes osteoporosis? Is it an estrogen deficiency? Do men have estrogen levels?

## Clinical Perspectives

1. How would you determine that the negative feedback system has failed for a particular hormone. What would you test for?

# REPRODUCTIVE SYSTEM  16

## CHAPTER OVERVIEW

The reproductive system contributes very little to the maintenance of homeostasis of the body. An individual can survive quite normally without reproducing. However, the survival of the species depends on the reproductive system. This is the major function of the system.

While the reproductive system is not a primary factor in overall homeostasis of the body there are unique situations in the reproductive process that involve homeostasis. These situations include Sertoli cells, endometrial cells, and the hormonal controls within the reproductive system.

The testes and ovaries produce the gametes with reduced chromosome numbers necessary for zygote formation through fusion of the egg and sperm. The reproductive system also produces hormones which play a role in initiation and completion of this reproductive function of the species. This body system is controlled by the central nervous system and the endocrine system. The various body systems maintain homeostasis in the body and cells of the body survive. But, if it were not for a pair of reproductive systems there would be no body.

## CHAPTER OUTLINE

### Introduction (Text page 532)

The reproductive system includes the gonads and reproductive tract (page 532).

The primary reproductive organs, or gonads, perform the dual function of (1) gametogenesis and (2) secreting sex hormones.

249

Testosterone in the male and estrogen in the female are responsible for the secondary sex characteristics.

The essential reproductive functions of the male are (1) spermatogenesis and (2) delivery of sperm to the female.

The essential female reproductive functions include (1) oogenesis; (2) reception of sperm; (3) fertilization; (4) gestation; (5) parturition; and (6) lactation.

### Reproductive cells each contain a half set of chromosomes (page 535).

Somatic cells contain forty-six chromosomes (the diploid number).

Gametes contain only one of each homologous pair for a total of twenty-three chromosomes (the haploid number).

### Gametogenesis is accomplished by meiosis (page 535).

During meiosis, a specialized diploid germ cell undergoes one chromosome replication followed by two nuclear divisions.

Genetic mixing provides novel new combinations of chromosomes.

### The sex of an individual is determined by the combination of sex chromosomes (page 535).

Twenty-two of the chromosome pairs are autosomal chromosomes.

The remaining pair of chromosomes are the sex chromosomes.

### Sex differentiation along male or female lines depends on the presence or absence of masculizing determinants during critical periods of embryonic development (page 535).

Gonadal specificity appears during the seventh week of intrauterine life.

The sex-determining region of the Y chromosome "masculinizes" the gonads.

By ten to twelve weeks of gestation, the sexes can be easily distinguished by the anatomical appearance of the external genitalia.

## Male Reproductive Physiology (Text page 537)

### The scrotal location of the testes provides a cooler environment essential for spermatogenesis (page 537).

In the last months of fetal life, the testes begin a slow descent, passing out of the abdominal cavity through the inguinal canal into the scrotum.

Descent of the testes into this cooler environment is essential, because spermatogenesis is temperature sensitive and cannot occur at normal body temperature.

### The testicular Leydig cells secrete masculinizing testosterone (page 537).

About eighty percent of the testicular mass consists of highly coiled seminiferous tubules, within which spermatogenesis takes place.

The endocrine cells that produce testosterone---the Leydig cells or interstitial cells---are located in the connective tissue between the seminiferous tubules.

Before birth, testosterone secretion by fetal testes is responsible for masculinizing the reproductive tract and external genitalia and for promoting descent of the testes into the scrotum.

At puberty, the Leydig cells start secreting testosterone, and spermatogenesis is initiated in the seminiferous tubules.

Testosterone is responsible for growth and maturation of the entire male reproductive system.

Ongoing testosterone secretion is essential for spermatogenesis and for maintaining a mature male reproductive tract throughout adulthood.

Testosterone is responsible for development of sexual libido at puberty and helps to maintain the sex drive in the adult male.

All male secondary sexual characteristics depend on testosterone for their development and maintenance.

Testosterone has a general protein anabolic effect and promotes bone growth.

Testosterone secretion and spermatogenesis occur continuously throughout the male's life.

## Spermatogenesis yields an abundance of highly specialized, mobile sperm (page 539).

Two functionally important cell types are present in the seminiferous tubules: germ cells and Sertoli cells.

During meiosis, each primary spermatocyte forms two secondary spermatocytes during the first meiotic division, finally yielding four spermatids as a result of the second meiotic division.

Sperm are essentially "stripped-down" cells in which most of the cytosol and the organelles not needed for the task of delivering the sperm's genetic information to an ovum have been extruded.

A spermatozoon has four parts: a head, an acrosome, a midpiece, and a tail.

## Throughout their development, sperm remain intimately associated with Sertoli cells (page 540).

The supportive Sertoli cells perform the following functions essential for spermatogenesis: (1) Sertoli cells provide nourishment for sperm cells. (2) Sertoli cells engulf cytoplasm extruded from spermatids during their remodeling and destroy defective germ cells that fail to successfully complete all stages of spermatogenesis. (3) The Sertoli cells secrete into the lumen seminiferous tubule fluid which flushes sperm into the epidymis. (4) Sertoli cells secrete androgen-binding protein. (5) The Sertoli cells are the site of action for control of spermatogenesis by both testosterone and FSH.

The Sertoli cells release inhibin, which acts in negative-feedback fashion to regulate FSH secretion.

## The two anterior pituitary gonadotropic hormones, LH and FSH, control testosterone secretion and spermatogenesis (page 541).

The testes are controlled by two gonadotropic hormones secreted by the anterior pituitary, luteinizing hormone (LH) and follicle-stimulating hormone (FSH).

Luteinizing hormone acts on the Leydig cells to regulate testosterone secretion.

Follicle-stimulating hormone acts on the seminiferous tubules, especially the Sertoli cells, to enhance spermatogenesis.

Secretion of both LH and FSH from the anterior pituitary is stimulated by a single hypothalamic hormone, gonadotropin-releasing hormone (GnRH).

Follicle-stimulating hormone is required for spermatid remodeling.

Gonadotropin-releasing hormone activity increases at puberty (page 542).

The pubertal process is initiated by an increase in GnRH activity sometime between eight and twelve years of age.

The factors responsible for initiating puberty in humans remain a mystery.

The ducts of the reproductive tract store and concentrate sperm and increase their motility and fertility as well (page 542).

The remainder of the male reproductive system consists of (1) a tortuous pathway of tubes that transport sperm from the testes to the outside of the body; (2) several glands, which contribute secretions that are important to the viability and motility of the sperm; and (3) the penis, which is designed to penetrate and deposit the sperm within the vagina of the female.

As they leave the testes, the sperm are incapable of either movement or fertilization.

Sperm gain both capabilities during their passage through the epididymis.

The epididymis concentrate the sperm by absorbing most of the fluid that enters from the seminiferous tubules.

The ductus deferens serves as an important site for sperm storage.

The accessory sex glands contribute the bulk of the semen (page 542).

The seminal vesicles: (1) supply fructose; (2) secrete prostaglandins; (3) provide more than half of the semen; and (4) secrete fibrinogen.

The prostate gland (1) secretes an alkaline fluid that neutralizes the acidic vaginal secretions, and (2) provides clotting enzymes and fibrinolysin.

The bulbourethral glands secrete a mucuslike substance that provides lubrication for sexual intercourse.

Prostaglandins are ubiquitous, locally acting chemical messengers (page 543).

Prostaglandins are produced in virtually all tissues from arachidonic acid, a fatty acid constituent of the phospholipids within the plasma membrane.

Besides enhancing sperm transport in semen, prostaglandins are known or suspected to exert other actions in the female reproductive system and in the respiratory, urinary, digestive, nervous and endocrine systems, in addition to having effects on platelet aggregation, fat metabolism, and inflammation.

Specific prostaglandins have been medically administered in such diverse situations as inducing labor, treating asthma, and treating gastric ulcers.

# Sexual Intercourse Between Males and Females (Text page 544)

The male sex act is characterized by erection and ejaculation (page 544).

The male sex act involves two components: (1) erection, and (2) ejaculation.

The sexual response cycle encompasses broader physiological responses that can be divided into four phases: (1) the excitement phase, (2) the plateau phase, (3) the orgasmic phase, and (4) the resolution phase.

Erection is caused by engorgement of the penis with blood.

During sexual arousal, the arterioles reflexly dilate and the erectile tissue fills with blood, causing the penis to enlarge both in length and width and to become more rigid.

Tactile stimulation of the glans reflexly triggers increased parasympathetic and decreased sympathetic activity to the penile arterioles.

This response is the major instance of direct parasympathetic control over blood-vessel caliber in the body.

Parasympathetic impulses promote secretion of lubricating mucus from the bulbourethral glands and the urethral glands in preparation for coitus.

The erection reflex is a spinal reflex that can be either facilitated or inhibited by higher brain centers through descending pathways that also terminate on the autonomic nerves supplying the penile arterioles.

Psychic stimuli can induce an erection.

Ejaculation is accomplished by a spinal reflex.

Sympathetic impulses cause sequential contraction of smooth muscles in the prostate, reproductive ducts, and seminal vesicles.

This contractile activity delivers prostatic fluid, then sperm, and finally seminal vesicle fluid into the urethra.

The filling of the urethra with semen triggers nerve impulses that activate a series of skeletal muscles at the base of the penis.

Rhythmic contractions of these muscles forcibly expel the semen through the urethra to the exterior.

During the resolution phase following orgasm, sympathetic vasoconstrictor impulses slow the inflow of blood into the penis so that the erection subsides.

### The female sexual cycle parallels that of the males in many ways (page 545).

Both sexes experience the same four phases of sexual cycle.

The excitement phase in females can be initiated by either physical or psychological stimuli.

Tactile stimulation of the clitoris and surrounding perineal area is an especially powerful stimulus.

Spinal reflexes bring about parasympathetically induced vasodilation of arterioles throughout the vagina and external genitalia.

Vasocongestion of the vaginal capillaries forces fluid out of the vessels into the vaginal lumen.

This fluid serves as the primary lubricant for intercourse.

Further vasocongestion of the lower third of the vagina reduces its inner capacity so that it tightens around the thrusting penis.

Simultaneously, the uterus raises upward, lifting the cervix and enlarging the upper two-thirds of the vagina to create a space for ejaculate deposition.

The sexual response culminates in orgasm as sympathetic impulses trigger rhythmic contractions of the pelvic musculature.

## Female Reproductive Physiology (Text page 546)

### Complex cycling characterizes female reproductive physiology (page 546).

The release of ova is intermittent, and secretion of female sex hormones displays wide cyclical swings.

During each menstrual cycle, the female reproductive tract is prepared for the fertilization and implantation of an ovum released from the ovary at ovulation.

The ovaries perform the dual function of oogenesis and secreting the female sex hormones, estrogen and progesterone.

Estrogen in the female is responsible for maturation and maintenance of the entire female reproduction system and establishment of female secondary sexual characteristics.

Estrogen is essential for ova maturation and release, development of physical characteristics that are sexually attractive to males, and transport of sperm from the vagina to the site of fertilization in the oviduct.

Estrogen contributes to breast development in anticipation of lactation.

Progesterone is important in preparing a suitable environment for nourishing a developing embryo/fetus and for contributing to the breast's ability to produce milk.

## Chromosome division in oogenesis parallels that in spermatogenesis, but there are major qualitative and quantitative sexual differences in gametogenesis (page 547).

The undifferentiated primordial germ cells in the fetal ovaries divide mitotically to give rise to millions of oogonia by the fifth month of gestation, at which time mitotic proliferation ceases.

Known now as primary oocytes, they contain forty-six replicated chromosomes.

Before birth, each primary oocyte is surrounded by a single layer of granulosa cells to form the primary follicle.

No new oocytes or follicles appear after birth.

A follicle is destined for one of two fates once it starts to develop: it will reach maturity and ovulate, or it will degenerate to form scar tissue.

By menopause, few, if any, primary follicles remain.

The primary oocyte within a primary follicle is still a diploid cell.

Just before ovulation, the primary oocyte, whose nucleus has been in meiotic arrest for years, completes its first meiotic division.

This division yields two daughter cells.

Almost all of the cytoplasm remains with one of the daughter cells, now called the secondary oocyte, which is destined to become the ovum.

Sperm entry into the secondary oocyte is needed to trigger the second meiotic division.

Twenty-three unpaired chromosomes remain in what is now the mature ovum.

These twenty-three maternal chromosomes unite with the twenty-three paternal chromosomes of the penetrating sperm to complete fertilization.

## The ovarian cycle consists of alternating follicular and luteal phases (page 548).

After the onset of puberty, the ovary constantly alternates between two phases---- the follicular phase, which is dominated by the presence of maturing follicles, and the luteal phase, which is characterized by the presence of the corpus luteum.

The follicle operates in the first half of the cycle to produce a mature egg ready for ovulation at midcycle.

The corpus luteum takes over during the second half of the cycle to prepare the female reproductive tract for pregnancy if fertilization of the released egg occurs.

The single layer of granulosa cells in a primary follicle proliferates to form several layers that surround the oocyte.

Follicular cells function as a unit to secrete estrogen.

The hormonal environment that exists during the follicular phase promotes enlargement and development of the follicular cell's secretory capacity, converting the primary follicle into a secondary, or antral, follicle capable of estrogen secretion.

This stage is characterized by the formation of a fluid-filled antrum.

The antrum occupies most of the space in a mature follicle.

The oocyte, surrounded by the zona pellucida and a single layer of granulosa cells, is displaced asymmetrically at one side of the growing follicle in a little mound that protrudes into the antrum.

The greatly expanded mature follicle bulges on the ovarian surface, creating a thin area that ruptures to release the oocyte at ovulation.

Just before ovulation, the oocyte completes its first meiotic division.

The ovum, still surrounded by its tightly adhering zona pellucida and granulosa cells, is swept out of the ruptured follicle into the abdominal cavity by the leaking antral fluid.

The released ovum is quickly drawn into the oviduct.

Rupture of the follicle at ovulation signals the end of the follicular phase.

Remnant follicular cells undergo dramatic structural transformations to form the corpus luteum, in a process called luteinization.

The follicular-turned luteal cells hypertrophy and are converted into very active steroid-ogenic tissue known as the corpus luteum.

Estrogen secretion in the follicular phase followed by progesterone secretion in the luteal phase is essential for preparing the uterus to be a suitable site for implantation of a fertilized ovum.

If the released ovum is not fertilized, the corpus luteum degenerates within fourteen days after its formation.

If fertilization and implantation do take place, the corpus luteum continues to grow and produce increasing quantities of progesterone and estrogen.

## The ovarian cycle is regulated by complex hormonal interactions among the hypothalamus, anterior pituitary, and ovarian endocrine units (page 550).

Gonadal function in the female is directly controlled by the anterior pituitary gonadotropic hormones, FSH and LH.

These hormones are regulated by hypothalamic gonadotropin-releasing hormone and feedback actions of gonadal hormones.

Antrum formation is induced by FSH.

Both FSH and estrogen stimulate proliferation of granulosa cells.

Both LH and FSH are required for synthesis and secretion of estrogen by the follicle.

Part of the estrogen produced by the growing follicle is secreted into the blood and is responsible for steadily increasing plasma estrogen levels during the follicular phase.

The secreted estrogen inhibits the hypothalamus and anterior pituitary in negative-feedback fashion.

The plasma FSH level declines during the follicular phase as the estrogen level rises.

The LH secretion continues to rise slowly during the follicular phase despite inhibition of GnRH secretion.

Ovulation and subsequent luteinization of the ruptured follicle are triggered by an abrupt, massive increase in LH secretion.

This LH surge brings about four major changes in the follicle: (1) It halts estrogen synthesis by the follicular cells. (2) It reinitiates meiosis in the oocyte of the developing follicle. (3) It triggers production of specific, locally acting prostaglandins. (4) It causes differentiation of follicular cells into the luteal cells.

The LH surge is triggered by a positive-feedback effect.

Thus, LH enhances estrogen production by the follicle, and the resultant peak estrogen concentration stimulates LH secretion.

Luteinizing hormone "maintains" the corpus luteum.

No progesterone is secreted during the follicle phase.

The follicular phase is dominated by estrogen and the luteal phase by progesterone. Even though a high level estrogen stimulates LH secretion, progesterone, which, dominates the luteal phase, powerfully inhibits LH secretion as well as FSH secretion. Inhibition of FSH and LH by progesterone prevents new follicular maturation and ovulation during the luteal phase.

Under progesterone's influence, the reproductive system is gearing up to support the just-released ovum, should it be fertilized.

## The uterine changes that occur during the menstrual cycle reflect hormonal changes during the ovarian cycle (page 553).

The menstrual cycle consists of three phases: the menstrual phase, the proliferative phase, and the secretory or progestational phase.

The menstrual phase is characterized by discharge of blood and endometrial debris from the vagina.

The first day of menstruation is considered to be the start of a new cycle and coincides with termination of the ovarian luteal phase and onset of the follicular phase.

As the corpus luteum degenerates, circulating levels of estrogen and progesterone drop precipitously.

The fall in ovarian hormone levels also stimulates release of a uterine prostaglandin that causes vasoconstriction of the endometrial vessels.

The entire uterine lining sloughs during each menstrual period except for a deep, thin layer of epithelial cells and glands from which the endometrium will regenerate.

Withdrawal of estrogen and progesterone upon degeneration of the corpus luteum leads simultaneously to sloughing of the endometrium and development of new follicles in the ovary under the influence of rising gonadotropic hormone levels.

The proliferative phase of the uterine cycle begins concurrent with the last portion of the ovarian follicular phase as the endometrium starts to repair.

The estrogen-dominant proliferative phase lasts from the end of menstruation to ovulation.

Following ovulation, a new corpus luteum is formed.

Progesterone acts on the thickened, estrogen-primed endometrium to convert it to a richly vascularized, glycogen-filled tissue.

This period is called the secretory phase, or the progestational phase.

If fertilization and implantation do not occur, the corpus luteum degenerates and a new follicular phase and menstruation phase begin once again.

## Fluctuating estrogen and progesterone levels produce cyclical changes in cervical mucus (page 554).

Under the influence of estrogen during the follicular phase, the mucus secreted by the cervix is abundant, clear, and thin.

After ovulation, under the influence of progesterone from the corpus luteum, the mucus becomes thick and sticky, essentially forming a plug across the cervical opening.

## Pubertal changes in females are similar to those in males, but menopausal changes are unique to females (page 554).

Unlike the fetal testes, the fetal ovaries do not need to be functional, because feminization of the female reproductive system automatically takes place in the absence of fetal testosterone secretion without the presence of female hormones.

The resultant secretion of estrogen by the activated ovaries induces growth and maturation of the female reproductive tract as well as development of the female secondary sexual characteristics.

Three other pubertal changes in females--growth of axillary and pubic hair, the pubertal growth spurt, and development of libido--are attributed to a spurt in adrenal androgen secretion at puberty, not to estrogen.

The pubertal rise in estrogen closes the epiphyseal plates.

The cessation of a women's menstrual cycles at menopause is attributable to the limited supply of ovarian follicles present at birth.

The termination of reproductive potential in a middle-aged woman is "preprogrammed" at her own birth.

## The oviduct is the site of fertilization (page 555).

Fertilization, the union of male and female gametes, normally occurs in the upper third of the oviduct.

The dilated end of the oviduct cups around the ovary and contains fimbriae, fingerlike projections that contract in a sweeping motion to guide the released ovum into the oviduct.

Fertilization must therefore occur within twenty-four hours after ovulation.

Sperm can survive about two days in the female reproductive tract.

Once sperm are deposited in the vagina at ejaculation, they must travel through the cervical canal, through the uterus, and then up to the egg in the upper third of the oviduct.

Once the sperm have entered the uterus, contractions of the myometrium churn them around in "washing-machine" fashion.

Sperm transport in the oviduct is facilitated by contractions of the oviduct smooth muscle.

These myometrial and oviduct contractions that facilitate sperm transport are induced by the high estrogen level that exists prior to ovulation, perhaps aided by seminal prostaglandins.

The acrosomal enzymes, which are exposed as the acrosomal membrane disrupts on contact with the corona radiata, enable the sperm to tunnel a path through these protective barriers.

The first sperm to reach the ovum itself fuses with the plasma membrane of the ovum, triggering a chemical change in the ovum's surrounding membrane that makes the outer layer impenetrable to the entry of any more sperm.

The head of the fused sperm is gradually pulled into the ovum's cytoplasm by a growing cone that engulfs it.

Penetration of the sperm into the cytoplasm triggers the final meiotic division of the secondary oocyte.

## The blastocyst implants in the endometrium through the action of its trophoblastic enzymes (page 557).

During the first six to seven days after ovulation, while the developing embryo is in transit in the oviduct and floating in the uterine lumen, the uterine lining is simultaneously being prepared for implantation under the influence of luteal-phase progesterone.

By the time the endometrium is suitable for implantation, the zygote has continued to proliferate and differentiate into a blastocyst capable of implantation.

Implantation begins when the trophoblastic cells overlying the inner cell mass release protein-digesting enzymes upon contact with the endometrium.

The trophoblast performs the dual functions of (1) accomplishing implantation as it carves out a hole in the endometrium for the blastocyst and (2) making metabolic fuel and raw materials available for the developing embryo as the advancing trophoblastic projections break down the nutrient-rich endometrial tissue.

After the blastocyst burrows into the decidua by means of trophoblastic activity, a layer of endometrial cells covers over the surface of the hole, completely burying the blastocyst within the uterine lining.

## The placenta is the organ of exchange between the maternal and fetal blood (page 559).

To sustain the growing embryo/fetus for the duration of its intrauterine life, the placenta is rapidly developed.

By day twelve, the trophoblastic layer is two cell-layers thick and is called the chorion.

Fingerlike projections of chorionic tissue extend into the pools of maternal blood.

The developing embryo sends out capillaries into these chorionic projections to form placental villi.

Each placental villus contains embryonic capillaries surrounded by a thin layer of chorionic tissue, which separates the embryonic/fetal blood from the maternal blood in the intervillus spaces.

The placenta is well established and operational by five weeks after implantation.

By this time, the heart of the developing embryo is pumping blood into the placental villi as well as to the embryonic tissues.

During intrauterine life, the placenta performs the functions of the digestive system, the respiratory system, and the kidneys for the "parasitic" fetus.

## Hormones secreted by the placenta play a critical role in the maintenance of pregnancy (page 561).

The placenta has the remarkable capacity to secrete a number of peptide and steroid hormones essential for the maintenance of pregnancy.

The most important are human chorionic gonadotropin, estrogen, and progesterone.

One of the first events following implantation is secretion by the developing chorion of human chorionic gonadotropin (hCG), a peptide hormone that prolongs the life span of the corpus luteum.

The maintenance of a normal pregnancy depends on high concentrations of progesterone and estrogen.

By the tenth week of pregnancy, hCG output declines to a low rate of secretion that is maintained for the duration of gestation.

The fall in hCG occurs at a time when the corpus luteum is no longer needed for its steroidal hormone output because the placenta has begun to secrete substantial quantities of estrogen and progesterone.

## Maternal body systems respond to the increased demands of gestation (page 562).

During gestation, a number of physical changes take place within the mother to accommodate the demands of pregnancy.

The most obvious change is uterine enlargement.

The breasts enlarge and develop the capability of producing milk.

The volume of blood increases and the cardiovascular system responds to the increasing demands of the growing placental mass.

Respiratory activity is increased.

Urinary output increases, and the kidneys excrete the additional wastes from the fetus.

## Parturition is accomplished by a positive feedback cycle (page 563).

Parturition requires (1) dilation of the cervical canal to accommodate passage of the fetus from the uterus through the vagina and to the outside and (2) contractions of the uterine myometrium that are sufficiently strong to expel the fetus.

Cervical softening is believed to be caused by relaxin, a peptide hormone produced by the placenta.

Relaxin also "relaxes" the birth canal by loosening the connective tissue between the pelvic bones.

Rhythmic, coordinated contractions begin at the onset of true labor.

The contractions occur with increasing frequency and intensity.

After having dilated the cervix sufficiently for passage of the fetus, these contractions force the fetus out through the birth canal.

A dramatic progressive increase in the concentration of myometrial receptors for oxytocin, probably induced by the increasing levels of estrogen during pregnancy, is ultimately responsible for initiating labor.

Oxytocin is a powerful uterine muscle stimulant and is known to play a role in the progression of labor.

Labor is divided into three stages: (1) cervical dilation, (2) delivery of the baby, and (3) delivery of the placenta.

After delivery, the uterus shrinks to its pregestational size, which is known as involution.

Involution occurs largely because of the precipitous fall in circulating estrogen and progesterone when the placental source of these steroids is lost at delivery.

## Lactation requires multiple hormonal inputs (page 567).

During gestation, the mammary glands, or breasts, are prepared for lactation.

The high concentration of estrogen promotes extensive duct development, whereas the high level of progesterone stimulates abundant alveolar-lobular formation.

Elevated concentrations of prolactin also contribute to mammary gland development by inducing the synthesis of enzymes needed for milk production.

The high estrogen and progesterone concentrations during the last half of pregnancy prevent lactation by blocking prolactins stimulatory action on milk secretion.

Prolactin is the primary stimulant of milk secretion.

Two hormones are critical for maintaining lactation: (1) prolactin and (2) oxytocin.

Release of these hormones is stimulated by a neuroendocrine reflex triggered by suckling.

Suckling of the breast by the infant stimulates sensory nerve endings in the nipple, resulting in initiation of action potentials that travel up the spinal cord to the hypothalamus.

The hypothalamus triggers a burst of oxytocin release from the posterior pituitary.

Oxytocin stimulates contractions of the myoepithelial cells in the breasts to bring about milk ejection.

Milk is composed of water, triglyceride fat, the carbohydrate lactose, a number of proteins, vitamins, and the minerals calcium and phosphate.

Colostrum, the milk produced for the first five days postpartum, contains lower concentrations of fat and lactose but higher concentrations of proteins, most notably lactoferrin and immunoglobulins.

The end is a new beginning (page 569).

> The single cell resulting from the union of male and female gametes divides mitotically and differentiates into a multicellular individual made up of a number of different organ systems that interact cooperatively to maintain homeostasis.
>
> All of the life-supporting homeostatic processes introduced throughout this book begin all over again at the start of a new life.

## KEY TERMS

Acrosome-p540
Autosomal chromosomes-p535
Blastocyst-p558
Chorion-p559
Colostrum-p569
Corpus luteum-p550
Cryptorchidism-p537
Decidua-p558
Follicular phase-p548

Gestation-p562
Homologous chromosomes-p535
Involution-p567
Leydig cells-p537
Luteal phase-p548
Luteinization-p550
Meiosis-p535
Oogonia-p547
Parturition-p563

Primary follicle-p547
Primary oocyte-p547
Primary spermatocytes-p539
Progestational phase-p554
Proliferative phase-p554
Secondary oocyte-p547
Seminiferous tubules-p537
Spermatids-p539
Trophoblast-p558

## REVIEW EXERCISES

**True - False** (Answers on page A-34)

_____ 1. Formation of the gonads in the embryo is known as gametogenesis.

_____ 2. The clitoris is a small erotic structure composed of tissue identical to the penis.

_____ 3. Fallopian tubes are the same as oviducts.

_____ 4. Somatic cells are haploid.

_____ 5. Gametes are haploid.

_____ 6. During meiosis, a diploid germ cell undergoes one chromosome replication followed by two nuclear divisions.

_____ 7. Gonadal specificity appears during the seventh week of intrauterine life.

_____ 8. SRY directs differentiation of the embryological gonad into testes.

_____ 9. Testosterone induces the male pattern of hair growth.

_____ 10. Sertoli cells secrete testosterone.

_____ 11. Each primary spermatocyte has twenty-three doubled chromosomes.

_____ 12. The acrosome is an enzyme-filled vesicle at the tip of the head of the sperm.

_____ 13. Sertoli cells phagocytize cytoplasm extruded from spermatids.

_____ 14. Seminiferous tubule fluid is secreted by Sertoli cells.

_____ 15. FSH acts on the Leydig cells to regulate testosterone secretion.

_____ 16. The bulbourethral glands secrete an alkaline fluid that neutralizes the acidic vaginal secretions.

_____ 17. Prostaglandins have been medically administered in treating gastric ulcers.

_____ 18. Increased parasympathetic activity to the penile arterioles produces an erection.

_____ 19. Progesterone in the female is responsible for maturation and maintenance of the entire female reproductive system.

_____ 20. The primary oocyte within a primary follicle is a haploid cell.

_____ 21. Sperm entry into the secondary oocyte is needed to trigger the second meiotic division.

_____ 22. The corpus luteum operates in the first half of the cycle to produce a mature egg ready for ovulation at midcycle.

_____ 23. Thecal and granulosa cells function as a unit to secrete estrogen.

_____ 24. Estrogen secretion in the follicular phase luteal phase followed by progesterone secretion in the luteal phase is essential for preparing the uterus for implantation.

_____ 25. Antrum formation is induced by LH.

_____ 26. The LH surge causes differentiation of follicular cells into luteal cells.

_____ 27. No progesterone is secreted during the follicular phase.

_____ 28. Under the influence of estrogen, the mucus secreted by the cervix is abundant, clear, and thin.

**Fill in the Blank** (Answers on page A-34)

29. In females, reproductive potential ceases during middle age at a time known as

_____.

30. Before birth, each primary oocyte is surrounded by a single layer of _____ to form a _____.

31. A follicle is destined for one of two fates once it starts to develop; it will reach maturity and ovulate, or it will degenerate to form scar tissue, a process known as

_____.

32. The thecal and granulosa cells, collectively known as _____, function as a unit to secrete estrogen.

33. The uterus consists of two layers: the _____ which is the outer smooth-muscle layer; and the _____ which is the inner lining that contains numerous blood vessels and glands.

34. Fertilization normally occurs in the _____.

35. The _____ (1) supply fructose, which serves as the primary energy source for ejaculated sperm; (2) secrete prostaglandins.

36. The _____ phase returns the genitalia and body systems to their prearousal state.

37. The penis consists almost entirely of _____ made up of three columns of spongelike vascular spaces.

38. Failure to achieve an erection in spite of appropriate stimulation is referred to as

_____.

39. Reproduction depends on the union of male and female _____.

40. _____ contain forty-six chromosomes.
41. Twenty-two of the chromosome pairs are _____.
42. When a testis remains undescended into adulthood, this condition is known as

    _____.
43. The endocrine cells that produce testosterone in males are _____
    or _____.
44. A spermatozoon has four parts: _____, _____, _____
    and _____.
45. The testes are controlled by the two gonadotropic hormones secreted by the anterior
    pituitary, _____, and _____.
46. A _____ is a single-layered sphere of cells encircling a fluid-filled cavity.
47. The endometrial tissue so modified at the implantation site is called the _____.
48. When the trophoblastic layer is two cell-layers thick it is then called _____.
49. _____ is a peptide hormone
    that prolongs the life span of the corpus luteum.
50. The period of gestation is about _____ weeks from conception.
51. _____ is a peptide hormones that is produced by the hypothalamus and is
    a key role in the progression of labor.
52. After delivery, the uterus shrinks to its pregestational size, a process known as

    _____.
53. _____ is the milk produced for the first five days postpartum.

**Matching:** Match the function to the placental hormone using the code on the right.
          Some functions may require multiple hormones (Answers on page A-35)
_____ 54. Helps prepare mammary glands for lactation     a. estrogen
_____ 55. Loosens connective tissue between pelvic        b. progesterone
          bones in preparation for parturition             c. hCG
_____ 56. Promotes formation of cervical mucus plug       d. relaxin
_____ 57. Maintains corpus luteum                         e. human chorionic
_____ 58. Softens cervix in preparation for cervical         gonadotropin
          dilation at parturition
_____ 59. Stimulates growth of myometrium, increasing
          uterine strength for parturition

**Multiple Choice** (Answers on page A-35)

_____ 60. Which of the following contraceptive practices, products, or methods is no longer popular due to reported complications including pelvic inflammatory disease and permanent infertility?

    a. birth-control pills

    b. morning after pills

    c. IUD's

    d. chemical spermides

    e. contraceptive sponge.

_____ 61. Which of the following contraceptive practices, products, or methods is used as a method of fertility?

    a. barrier method

    b. coitus interruptus

    c. oral contraceptives

    d. rhythm method

    e. none of the above

_____ 62. Which of the following devices used as barriers is the easiest to use?

    a. condom

    b. diaphragm

    c. cervical cap

    d. condomlike contraceptive for women

    e. vaginal cap

_____ 63. Which of the following contraceptive practices, products, or methods the risk of intravascular clotting, especially in women who also smoke tobacco?

    a. IUD's

    b. morning after pill

    c. foams

    d. implanted contraceptives

    e. contraceptive sponge

_____ 64. Which of the following contraceptive practices, products, or methods should be used only once due to increased risk of cardiovascular disease?

    a. sterilization

    b. birth-control pills

    c. IUD's

    d. morning after pill

    e. subcutaneous implantation of synthetic progesterone

_____ 65. Which is the longest acting method of contraception?
      a. diaphragms
      b. IUD's
      c. morning after pills
      d. birth-control pills
      e. subcutaneous implantation of synthetic progesterone

_____ 66. Which of the following contraceptive techniques has the highest failure rate?
      a. implanted contraceptives
      b. barrier methods
      c. rhythm method
      d. intrauterine device
      e. oral contraceptives

_____ 67. Which of the following hormones is the primary stimulant of milk secretion?
      a. human chorionic somatomammotropin
      b. prolactin
      c. estrogen
      d. progesterone
      e. prostaglandins

_____ 68. Which of the following hormones promotes fat deposition?
      a. prolactin
      b. human chorionic somatomammotropin
      c. estrogen
      d. progesterone
      e. none of the above

_____ 69. Which of the following hormones secreted by the corpus luteum softens the cervix?
      a. estrogen
      b. progesterone
      c. oxytocin
      d. relaxin
      e. none of the above

**Point to Ponder**

1. It is very commonplace to discuss abnormalities in hormone secretion of all endocrine glands and tissues except the gonads. Is it because such abnormalities concerning the gonads do not exist or is it due to a social attitude?

2. After carefully studying this chapter, do you think gay people could have abnormal hormone levels?

3. Why do you think most oral contraceptives are for women? Is there a physiological reason?

**Clinical Perspectives**

1. Explain why birth control pills are prescribed to women who have complications at the onset of menopause.
2. Explain how an infant girl's breasts can begin to enlarge.

# ANSWERS to Chapter 1 (pages 1 - 8)

**True - False**
1. True
2. False - its reversed
3. False - some organs are in several systems
4. True
5. False - Extrinsic control
6. True
7. True
8. True
9. True
10. False - The internal environment must be maintained in a dynamic steady state instead of a fixed state. Fluctuations around an optimal level are minimized by compensatory physiological responses so that the homeostatically maintained factors are kept within the narrow limits that are compatible with life.
11. True
12. False - Exocrine glands
13. False - Intrinsic and extrinsic
14. True
15. True

**Fill in the Blank**
16. Exocrine
17. Lumen
18. Homeostasis
19. Interstitual
20. Digestive
21. Negative Feedback
22. Cell
23. Internal environment
24. Extracellular fluid, Plasma, Interstitial fluid
25. Negative Feedback
26. Pathophysiology
27. Cell
28. E.G.F. (Epidermal Growth Factor)
29. Glands
30. Body systems
31. Circulatory, Digestive, Respiratory, Urinary, Skeletal, Muscular, Integumentary, Immune, Nervous, Endocrine, and Reproductive.

**Matching**

32. E
33. G
34. K
35. H
36. J
37. A
38. D
39. F
40. B
41. I
42. C

**Multiple Choice**

43. C
44. D
45. C
46. B
47. C
48. A
49. B
50. A
51. C

52. A
53. E
54. A
55. D
56. B
57. A
58. B
59. B

# ANSWERS to Chapter 2 (pages 9 - 20)

## True - False

1. False - accumulation of gangliosides
2. False - occurs in inner mitochondrial membrane
3. True
4. False - glycolysis is anaerobic
5. False - another name for electron transport chain
6. True
7. True
8. True
9. False - plant cells, not human or other animal cells, are surrounded by a cell wall
10. True
11. True
12. True
13. False - single cell
14. False - DNA
15. True
16. False - ribosomes
17. True
18. False - ribosomes found only on the rough ER
19. True
20. True
21. False - Golgi sacs
22. True
23. True
24. False - double membrane
25. False - ATP
26. False - 2 molecules of ATP
27. True
28. True
29. True
30. True
31. True
32. False - Microtubules are found in cilia
33. False - Secretory vesicles release their contents to the outside
34. False - Peroxisomes contain oxidative enzymes
35. False - Glycolysis occurs in the cytosol.
36. True
37. False - Pyruvic acid is the end product of glycolysis

## Fill in the Blank

38. messenger
39. transport vesicles
40. microtrabecular lattice
41. ribosomal
42. Lysosomes
43. basil body
44. peroxisomes
45. plasma membrane, nucleus, cytoplasm
46. cytoskeleton
47. proteins, lipids
48. vaults
49. smooth
50. exocytosis
51. hydrolytic
52. endocytosis
53. Microfilaments
54. DNA
55. Microtubules
56. plasma membrane

## Matching

57. G
58. K
59. J
60. A
61. I
62. H
63. B
64. C
65. F
66. E
67. D

## Multiple Choice

| | |
|---|---|
| 68. B | 78. B |
| 69. E | 79. B |
| 70. A | 80. D |
| 71. B | 81. B |
| 72. A | 82. B |
| 73. D | 83. A |
| 74. B | 84. B |
| 75. A | 85. B |
| 76. E | 86. C |
| 77. A | 87. B |

# ANSWERS to Chapter 3 (pages 21 - 39)

**True - False**

1. True
2. False - phagocytosis
3. False - there is no net movement; at equilibrium an equal number of molecules are moving in opposite directions across the membrane
4. False - a potential of 70mV is of greater magnitude than a potential of 30mV. The sign refers to the charge on the inside of the cell.
5. True
6. True
7. True
8. False - outer surface
9. True
10. False - is possible
11. False - channel regulation mechanism
12. True
13. False - collagen, elastin, fibronectin
14. True
15. False - fibroblast instead of fibronectin
16. False - tight junction
17. True
18. False - both charged and uncharged molecules are soluble in water. The size of the particle is the second property that influences permeability
19. False - 8
20. False - the faster the rate of net diffusion
21. True
22. True
23. False - the slower the rate of diffusion
24. True
25. True
26. True
27. False - opposite charges not like charges
28. True
29. False - synapses operate in one direction only
30. False - a given synapse is either always excitatory or always inhibitory
31. False - larger the graded potential
32. False - repolarization
33. True
34. False - -50mV to -55mV is the threshold potential
35. True
36. True
37. True

38. True
39. False - 50 times
40. False - urgent information is carried by myelinated fibers, less urgent by unmyelinated
41. True
42. False - lowest
43. False - deficiency
44. True

**Fill in the Blank**
45. cystic fibrosis
46. second messenger
47. first messenger
48. phosphorylation
49. extracellular matrix
50. cell adhesion molecules in the cells plasma membranes, the extracellular matrix, and specialized cell junctions
51. collagen, elastin, fibronectin
52. fibroblasts
53. desmosomes, tight junctions, gap junctions
54. tight junctions
55. collagen
56. membrane potential, negative
57. phospholipids
58. hydrophilic, hydrophobic
59. cholesterol
60. carrier molecules, receptor sites
61. elastin
62. gap junctions, connexons
63. fibronectin
64. heart, smooth
65. net diffusion
66. state of equilibrium, steady state
67. impermeable
68. carrier-mediated transport
69. transport maximum
70. Primary active transport
71. endocytosis
72. phagocytosis
73. membrane potential
74. cations, negatively, anions, positively
75. electrical gradient
76. electrochemical gradient
77. concentration

78. osmosis
79. osmotic pressure
80. resting membrane potential
81. Dendrites, axon, nerve fibers
82. nodes of Ranvier
83. polarization
84. saltatory conduction
85. synaptic cleft
86. 0.5 to 14 sec
87. voltage-gated, chemical messenger-gated, mechanically-gated
88. axon hillock
89. Glycine, Glutamata, Gamma-aminobutyric acid (GABA)
90. spike
91. cell body, dendrites
92. lipids
93. all-or-none law
94. subsynaptic membrane
95. divergence

| **Matching** | **Multiple Choice** | |
|---|---|---|
| 96. D | 106. C | 116. B |
| 97. E | 107. B | 117. D |
| 98. A | 108. C | 118. C |
| 99. D | 109. D | 119. E |
| 100. D | 110. A | |
| 101. A | 111. D | |
| 102. B | 112. B | |
| 103. D | 113. B | |
| 104. C | 114. D | |
| 105. A | 115. C | |

# ANSWERS to Chapter 4 (pages 41 - 57)

**True -False**

1. False - smallest
2. True
3. False - opposite
4. True
5. False - left
6. True
7. True
8. False - afferent
9. True
10. False - afferent neuron
11. False - efferent neuron
12. True
13. True
14. True
15. True
16. False - cerebrocerebellum
17. False - cerebellum
18. False - glial cells or neuroglia
19. True
20. False - oligodendrocytes
21. True
22. True
23. True
24. False - cannot
25. True
26. False - 18 inches
27. False - five lumbar nerves
28. False - descending tracts
29. False - afferent
30. True
31. True
32. True
33. True
34. False - decreasing
35. False - activity is not reduced
36. False - twenty percent
37. False - paradoxical
38. True
39. False - basal nuclei
40. True

41. False - long-term memory
42. True
43. True
44. True
45. False - rapid
46. True
47. False - nervous system
48. True
49. False - synaptic cleft

## Fill in the Blanks
50. spinocerebellum
51. vestibulocerebellum
52. cerebrocerebellum
53. simple (basic reflexes) and acquired (conditioned reflexes)
54. cervical, thoracic, lumbar, sacral, coccygeal
55. caudal equina
56. afferent, efferent
57. ganglion, center or a nucleus
58. spinal nerve
59. nerve
60. hypothalamus, thalamus
61. cerebellum
62. cerebral cortex
63. grey, white
64. parietal lobes, somesthetic sensations
65. Broca'a, frontal lobe, Wernick's, cortex
66. glial cells, neuroglia
67. Astrocytes
68. Oligodendrocytes
69. Microglia
70. medulla, pons, midbrain
71. reticular formation
72. consciousness
73. coma
74. paradoxical
75. slow-wave
76. REM (paradoxical)
77. basal nuclei, thalamus, hypothalamus
78. thalamus
79. hypothalamus
80. motivation
81. Learning
82. consolidation

83. brain, spinal cord, peripheral nervous system
84. afferent division
85. afferent neurons, efferent neurons, interneurons
86. CNS

**Matching**

87. B
88. D
89. A
90. A
91. A
92. A
93 C
94. A
95. B
96. A
97. A

**Multiple Choice**

98. C
99. C
100. B
101. D
102. B
103. C
104. A
105. E
106. D
107. C

**True - False**

1. False - none of the nociceptors
2. True
3. False - unmyelinated fibers
4. True
5. False - substance P
6. False - afferent fibers
7. True
8. True
9. False - sour
10. True
11. True
12. True
13. True
14. True
15. False - phasic receptors
16. False - it can
17. True
18. True
19. False - vitreous humor
20. True
21. True
22. True
23. False - rod shaped
24. False - other media's, such as water
25. True
26. False - pinna
27. True
28. False - three
29. False - endolymph
30. False - basilar membrane
31. True
32. False - cones
33. True
34. False - cells do not touch neuromuscular joint
35. True
36. False - opening not closing
37. True
38. False - just reversed
39. False - turned off
40. True

41. True
42. True
43. False - somatic nervous system
44. True
45. False - long
46. False - parasympathetic
47. True
48. True
49. True
50. False - sympathetic

**Fill in the Blank**
51. sclera
52. quality
53. accommodation
54. cataract
55. optic disc
56. opsin
57. optic tracts
58. taste pore
59. cortical gustatory area
60. olfactory nerve
61. taste buds
62. subcortical and thalamic-cortical
63. olfactory receptors, supporting cells, basal cells
64. sensory input
65. phasic
66. glaucoma
67. sensory input
68. phantom pain
69. vestibular apparatus
70. cerumen
71. auditory nerve
72. basilar
73. otolith organs
74. mechanical, thermal, polymodal
75. A-delta fibers
76. substance P
77. prostaglandins
78. opiate receptors
79. placebo
80. motor neurons
81. final common pathway
82. neuromuscular junction

83. muscle fiber
84. terminal button
85. motor end plate
86. ACh
87. Botulinum toxin
88. Organophosphates
89. myasthenia gravis
90. postganglionic fiber, preganglionic fiber
91. adrenergic fibers
92. varicosities
93. epinephrine (adrenaline)
94. Nicotinic receptors
95. Muscarinic receptors

| **Matching** | | **Multiple Choice** | |
|---|---|---|---|
| 96. G | 106. A | 116. C | 126. A |
| 97. A | 107. B | 117. E | 127. C |
| 98. J | 108. B | 118. A | 128. D |
| 99. B | 109. A | 119. D | 129. A |
| 100. C | 110. A | 120. D | 130. A |
| 101. I | 111. B | 121. A | 131. B |
| 102. D | 112. B | 122. E | 132. D |
| 103. H | 113. B | 123. B | 133. E |
| 104. F | 114. A | 124. B | 134. A |
| 105. E | 115. B | 125. D | 135. A |

# ANSWERS to Chapter 6 (pages 81 - 95)

**True - False**

1. True
2. False - intrafusal
3. True
4. False - negative feedback
5. False - externally applied tension
6. True
7. True
8. False - single nucleus
9. True
10. True
11. False - calcium concentration
12. False - myosin ATPase
13. True
14. False - myoglobin
15. False - pyruvic acid
16. False - oxidative fibers
17. False - muscle fiber
18. True
19. True
20. False - actin spiral
21. True
22. True
23. False - calcium binds
24. True
25. True
26. False - connective tissue
27. False - produces sustained contractions
28. True
29. False - motor-unit recruitment
30. True
31. False - origin
32. False - lengthens
33. True
34. False - sarcoplasmic reticulum
35. False - lateral sacs
36. True
37. False - latent period
38. True
39. False - calcium

## Fill in the Blank

40. Excitation-contraction coupling
41. transverse tubule
42. sarcoplasmic reticulum
43. lateral sacs
44. actin binding, ATPase
45. rigor mortis
46. latent period
47. Multiunit smooth-muscle
48. Neurogenic
49. functional syncytium
50. Slow-wave potentials
51. Latch-phenomenon
52. muscle fiber
53. myofibrils
54. A band
55. sarcomere
56. myosin, actin
57. Tropomyosin
58. Tropomyosin, troponin
59. gamma motor neuron, alpha motor neurons
60. Golgi tendon organs
61. motor unit
62. tension
63. series-elastic
64. origin, insertion
65. isotonic
66. concentric, eccentric
67. Creatine phosphate
68. Oxidative phosphorylation
69. myoglobin
70. Disuse
71. Muscular dystrophy

## Matching

| | |
|---|---|
| 72. B | 81. B |
| 73. A | 82. A |
| 74. C | 83. C |
| 75. A | |
| 76. C | |
| 77. A | |
| 78. C | |
| 79. A | |
| 80. B | |

## Multiple Choice

| | |
|---|---|
| 84. C | 93. C |
| 85. A | |
| 86. A | |
| 87. E | |
| 88. D | |
| 89. B | |
| 90. B | |
| 91. E | |
| 92. B | |

**True - False**
1. False - Contractile cells
2. False - SA node
3. True
4. True
5. True
6. False - long
7. True
8. False - P wave
9. True
10. False - decrease
11. False - sympathetic
12. True
13. True
14. False - LDL
15. True
16. True
17. True
18. False - Late
19. True
20. True
21. True
22. False - produces no sound
23. True
24. True
25. True
26. True
27. False - systemic circulation
28. False - Function as two separate pumps
29. False - right side
30. True
31. False - pulmonary circulation
32. True
33. True

**Fill in the Blank**
34. Vascular spasm
35. Atherosclerosis
36. thrombus, embolus
37. Collateral circulation
38. HDL, LDL, VLDL

39. fibrillation
40. interatrial
41. T wave
42. Tachycardia
43. bradycardia
44. apex
45. cardiopulmonary resuscitation
46. Ventricles
47. veins
48. aorta
49. tricuspid valve
50. epicardium
51. latent pacemakers
52. Cardiac output
53. Cardiac reserve
54. intrinsic control
55. Frank-Starling Law of the Heart
56. Heart failure
57. end-diastolic volume
58. isovolumetric ventricular contraction
59. end-systolic volume (ESV)
60. murmurs
61. systolic murmur

## Matching
62. B
63. A
64. B
65. B
66. A
67. A
68. A
69. A
70. B
71. A

## Multiple Choice
72. E
73. D
74. B
75. C
76. B
77. C
78. B
79. C
80. D
81. A

# ANSWERS to Chapter 8 (pages 113 - 126)

**True - False**
1. True
2. True
3. False - vary from organ to organ
4. False - exchanges are not made directly
5. True
6. False - is a function of the lymphatic system
7. False - right
8. True
9. False - the greater the rate of flow
10. False - does not fluctuate
11. True
12. True
13. True
14. False - Transient hypotension
15. True
16. False - short term
17. True
18. False - unchanged
19. False - does not respond and adapts or "reset" to operate at a higher level
20. True
21. True
22. False - can either be chemical or physical
23. True
24. True
25. False - except the brain
26. True
27. True
28. False - decreasing
29. True
30. False - can only be driven forward
31. True
32. False - 5mm Hg

**Fill in the Blank**
33. Hypertension
34. Hypotension
35. Orthostatic hypertension
36. Irreversible shock
37. Capacitance vessels
38. Venous capacity

39. Venous return
40. Varicose veins
41. Baroreceptors
42. Carotid-sinus, aortic-arch baroreceptors
43. Arteries
44. Flow rate
45. 120 mm Hg
46. pulse pressure
47. Mean arterial pressure
48. arterioles
49. diffusion
50. Bulk flow
51. Capillary blood pressure
52. lymphatic system
53. edema
54. Vasoconstriction
55. Vascular tone
56. Endothelial-derived relaxing factor (EDRF)
57. Endothelial cells
58. cardiovascular control center

**Matching**

59. A
60. A
61. A
62. B
63. C
64. A
65. B
66. A
67. B
68. A
69. A
70. A
71. B
72. B

**Multiple Choice**

73. C
74. D
75. A
76. E
77. C
78. B
79. C
80. E
81. B
82. E

# ANSWERS to Chapter 9 (pages 127 - 149)

**True - False**

1. True
2. True
3. False - trillion
4. True
5. False - Pernicious anemia
6. False - Primary
7. False - thrombin
8. True
9. False - seven steps
10. True
11. True
12. True
13. False - vitamin K
14. False - 90%
15. True
16. False - sodium and chloride
17. True
18. True
19. True
20. True
21. False - partially responsible for changes
22. True
23. False - red dye
24. False - an oval or kidney-shaped nucleus
25. True
26. True
27. True
28. False - histamine
29. False - B lymphocytes
30. False - B lymphocytes
31. True
32. False - Two.  Antibody-mediated immunity and cell-mediated immunity
33. True
34. True
35. True
36. False - 5
37. True
38. True
39. False - no naturally occurring antibodies
40. True
41. True

42. False - autoimmune responses
43. False - 20 minutes
44. True
45. True
46. True
47. False - helper T cells
48. True
49. False - T cells
50. False - do not result in malignancy
51. True
52. False - has no direct blood supply
53. True
54. False - are able to convert
55. False - does have control
56. True
57. False - B lymphocytes
58. False - Interferon
59. True
60. True
61. True

**Fill in the Blank**
62. Hemostasis
63. Blood coagulation, clotting
64. tissue thromboplastin
65. Serum
66. plasmin
67. emboli
68. hemophilia
69. spleen
70. erythropoiesis
71. Aplastic
72. Anemia
73. Hemolytic
74. Sickle-cell anemia
75. nutrients, waste products, dissolved gases, hormones
76. plasma proteins
77. albumins, globulins, fibrinogen
78. Albumins
79. Leukocytes
80. polymorphonuclear granulocytes
81. Lymphocytes
82. phagocytosis
83. Neutrophils
84. Basophils

85. cell-mediated immune response
86. Monocytes
87. gamma globulins, immunoglobulins
88. Agglutination
89. complement system, phagocytes, killer cells
90. B cells, antigen
91. active immunity
92. interleukin 1
93. cell-mediated immunity
94. Cytotoxic T-cells
95. perforin
96. interleukin 2
97. Helper T cells
98. Myasthenia gravis
99. AIDS
100. allergy, hypersensitivity
101. anaphylactic shock
102. Immunity
103. Lymphoid tissue
104. histamine
105. fibrinogen, fibrin
106. salicylates, glucocorticoids
107. antibody-mediated, cell-mediated
108. thymus, thymosin
109. Haptens
110. immediate hypersensitivity
111. spot desmosomes
112. sebaceous glands, sebum
113. sweat glands, the sweat pores
114. Melanocytes, melanin
115. keratinocytes
116. Langerhans cells
117. Granstein cells
118. alveolar macrophages

**Matching**

| | | | **Multiple Choice** | | |
|---|---|---|---|---|---|
| 119. H | 127. G | 135. B | 140. B | 148. D | 156. A |
| 120. A | 128. E | 136. C | 141. D | 149. B | 157. B |
| 121. J | 129. B | 137. D | 142. B | 150. E | 158. E |
| 122. B | 130. C | 138. D | 143. C | 151. C | 159. A |
| 123. C | 131. B | 139. B | 144. E | 152. D | |
| 124. I | 132. A | | 145. E | 153. B | |
| 125. D | 133. D | | 146. E | 154. E | |
| 126. F | 134. A | | 147. A | 155. B | |

# ANSWERS to Chapter 10 (pages 151 - 167)

**True - False**
1. True
2. False - when compliance is decreased
3. False - 5.7 liters
4. True
5. True
6. True
7. False - 100 mmHg
8. True
9. True
10. True
11. False - between the alveoli and the blood
12. True
13. False - DRG
14. True
15. True
16. True
17. False - increase
18. True
19. False - 1.5 percent
20. True
21. False - decreased
22. False - 97.5 % saturated
23. True
24. True
25. False - circulatory hypoxia
26. True
27. False - higher pressure to lower pressure
28. False - lower
29. True
30. False - decreases
31. False - increases
32. True
33. True

**Fill in the Blank**
34. Histotoxic hypoxia
35. Hypercapnia
36. oxygen-hemoglobin dissociation curve
37. Bohr effect
38. carboxyhemoglobin

39. Haldane effect
40. Hypoxic hypoxia
41. Pneumonia
42. partial pressure
43. partial pressure gradient
44. 60 mm Hg, 6 mm Hg
45. Functional residual capacity
46. obstructive, restrictive
47. anatomic dead space
48. Atmospheric (barometric) pressure
49. Bronchodilation
50. Emphysema
51. Compliance
52. Tidal volume
53. medullary respiratory center
54. apneustic center, pneumotaxic center
55. dorsal respiratory group
56. Apneusis
57. carotid bodies, aortic bodies
58. Apnea
59. dyspnea
60. Respiration
61. trachea, esophagus
62. Vocal cords
63. Type I alveolar cells
64. Type II alveolar cells

**Matching**

65. A
66. F
67. E
68. A
69 G
70. D
71. B
72. H
73. C
74. D
75. G
76. C

**Multiple Choice**

77. C
78. E
79. A
80. B
81. E
82. C
83. E
84. B
85. A
86. D
87. B
88. A

# ANSWERS to Chapter 11 (pages 169 - 181)

**True - False**

1. True
2. False - arterial blood
3. True
4. False - cortex
5. False - glomerular membrane
6. False - and the filtration coefficient
7. True
8. False - cardiac artria
9. False - renal threshold
10. True
11. False - proximal tubule
12. True
13. False - actively
14. False - hydrogen and potassium
15. True
16. True
17. False - 1.5 liters
18. False - less
19. False - greater
20. True
21. False - hypothalamus

**Fill in the Blank**

22. Urea
23. sodium
24. phosphate and calcium
25. expands
26. renin-angiotensin-aldosterone
27. angiotensin I
28. angiotensin II
29. Aldosterone
30. transepithelial
31. mesangial cells
32. filtration coefficient, net filtration pressure
33. proximal
34. osmotic
35. impermeable
36. Peristaltic
37. internal urethral sphincter
38. 300

39. sodium-potassium pump
40. potassium secretion

**Matching**

41. A
42. B,C
43. A,B,C
44. A
45. B
46. A
47. A
48. A
49. B
50. A
51. A

**Multiple Choice**

52. E
53. C
54. B
55. D
56. A
57. B
58. D
59. B
60. A
61. E

# ANSWERS to Chapter 12 (pages 183 - 193)

## True - False
1. False - positive balance
2. False - two-thirds
3. False - extracellular fluid
4. True
5. True
6. True
7. True
8. False - acids
9. True
10. True
11. False - chemical buffer system
12. False - reversed, ammonia combines with hydrogen to form ammonium ions
13. False - diabetes mellitus
14. True

## Fill in the Blank
15. stable balance
16. Plasma, interstitial fluid, interstitial fluid
17. obligatory loss of salt in sweat, feces, urine
18. renin-angiotensin-aldosterone
19. insufficient water intake, excessive water loss, diabetes insipidus
20. hypothalamic osmoreceptors
21. pH
22. phosphate buffer system
23. hypoventilation
24. metabolic acidosis
25. metabolic alkalosis
26. hydrogen
27. hydrogen excretion, bicarbonate excretion, ammonia secretion

## Matching
28. C
29. D
30. A
31. A
32. C
33. C
34. B
35. C

## Multiple Choice
36. E
37. B
38. C
39. E
40. C
41. E
42. D
43. C
44. B
45. B

# ANSWERS to Chapter 13 (pages 195 - 213)

## True - False

1. False - monosaccharides not disaccharides
2. False - polysaccharides not disaccharides
3. False - monoglycerides not triglycerides
4. True
5. False - mucosa not submucosa
6. False - Auerbach's not Haustra's
7. True
8. False - skeletal not smooth
9. True
10. True
11. True
12. True
13. False - eructation not borborygmia
14. False - fundus not antrum
15. True
16. True
17. True
18. False - vomiting or emesis not malocclusion
19. False - pepsinogen not mucus
20. False - parietal cells not mucous neck cells
21. True
22. False - pepsinogen not trypsinogen
23. True
24. True
25. False - water is not absorbed from the stomach contents
26. False - acini not isles of Langerhans
27. False - Enterokinase not HCI
28. False - alkaline not acidic
29. True
30. True
31. True
32. False - liver not gallbladder
33. True
34. True
35. False - microvilli are known as brush border
36. True
37. True
38. True
39. True

## Fill in the Blank

40. succus entericus
41. Enterokinase
42. Aminopeptidases
43. villi
44. Malabsorption
45. crypts of Lieberkuhn
46. mastication
47. 99.5%, 0.5%
48. Xerostomia
49. acquired or conditioned, salivary reflex
50. palate
51. nitroglycerin
52. bolus
53. oropharyngeal stage
54. gastroesophageal sphincter
55. Peristalsis
56. Secondary peristaltic waves
57. appendix
58. Haustral contractions
59. colon, cecum, appendix, rectum
60. GIP (gastric inhibitory peptide)
61. fundus
62. antrum
63. gastric pits
64. gastrin
65. zymogen granules
66. Gastric
67. peptic ulcer
68. Steatorrhea
69. liver
70. sphincter of Oddi
71. Bile salts
72. choleretic
73. CCK
74. hydrolysis

## Matching

| | |
|---|---|
| 75. A | 80. M |
| 76. D | 81. H |
| 77. G | 82. K |
| 78. L | 83. F |
| 79. B | |

## Multiple Choice

| | |
|---|---|
| 84. C | 89. C |
| 85. B | 90. E |
| 86. E | 91. B |
| 87. A | 92. D |
| 88. E | |

# ANSWERS to Chapter 14 (pages 215 - 222)

**True - False**
1. False - external work
2. False - energy can be converted to another form
3. True
4. False - hypothalamus
5. True
6. True
7. True
8. True
9. False - normally rectal temperature is a degree higher
10. True
11. False - conduction uses direct contact
12. True
13. True
14. True

**Fill in the Blank**
15. Increased
16. epinephrine, thyroid hormone
17. endogenous pyrogen
18. anterior, decreasing, vasodilation
19. prostaglandins
20. sustained exercise
21. sympathetic
22. Convection
23. lack of appetite
24. lipostatic
25. external work, internal work
26. Positive
27. renal failure, cancer, tuberculosis
28. Radiation

**Matching**
29. D
30. A
31. F
32. B
33. C
34. E

**Multiple Choice**
35. C
36. E
37. B
38. D
39. C
40. C
41. B
42. D
43. C
44. D

# ANSWERS to Chapter 15 (pages 223 - 247)

**True - False**
1. True
2. True
3. True
4. True
5. False - insulin
6. False - estrogens
7. True
8. False - neurohypophysis
9. False - vasopressin
10. True
11. True
12. False - non-tropic hormone
13. False - several not just one
14. False - lipophilic
15. True
16. False - lipophilic
17. True
18 True
19. False - $T_3$
20. True
21. False - increase
22. False - cretinism
23. True
24. True
25. False - eighty percent
26. True
27. False - especially muscle
28. True
29. False - primary not secondary
30. True
31. False - four hours
32. True
33. True
34. True
35. False - Type I
36. True
37. True
38. False - increase
39. True
40. True

41. False - decreases
42. True
43. True
44. False - increases
45. True

**Fill in the Blank**
46. gonadotropins
47. pituitary gland
48. Androgens
49. osteoblasts
50. Cartilage
51. diaphysis, epiphysis
52. dwarfism
53. gigantism
54. Hormones
55. Growth-hormone-releasing-hormone
56. Melanocyte-stimulating hormone, melanin
57. hyperplasia, hypertrophy
58. endocrine system
59. HRE (Hormone Response Element)
60. tropic
61. steroids
62. diurnal, circadian rhythm
63. permissiveness
64. Synergism
65. Antagonism
66. hypophysiotropic hormones
67. hypothalamus, anterior pituitary
68. amines
69. acromegaly
70. Metabolism
71. Anabolism
72. Catabolism
73. islets of Langerhans
74. Glycogenesis
75. low
76. Calcitonin
77. cholecalciferol
78. rickets, osteomalacia
79. osteoporosis
80. thyroglobulin
81. Grave's disease
82. goiter

83. C-cells
84. monoiodotyrosine
85. thyroxine
86. zona glomerulosa, zona fasciculata, zona reticularis
87. Cortisol
88. Cushings
89. Hirsutism
90. Addison's
91. chromaffin granules
92. pheochromocytoma

**Matching**

| | | |
|---|---|---|
| 93. S | 119. A | |
| 94. B | 120. N | |
| 95. K | 121. G | |
| 96. R,J | 122. N | |
| 97. D | 123. S | |
| 98. C | 124. N | |
| 99. R | 125. B | |
| 100. R | 126. E | |
| 101. L | 127. A | |
| 102. N | 128. H | |
| 103. E | 129. G | |
| 104. N | 130. E | |
| 105. Q | 131. E | |
| 106. J | 132. H | |
| 107. O | 133. B | |
| 108. I | 134. D | |
| 109. N | 135. E | |
| 110. E | 136. B | |
| 111. F | 137. A | |
| 112. S | 138. C | |
| 113. J,R | 139. C | |
| 114. P | 140. E | |
| 115. C | 141. B | |
| 116. M | 142. A | |
| 117. N | 143. D | |
| 118. E | | |

**Multiple Choice**

| | |
|---|---|
| 144. C | |
| 145. B | |
| 146. E | |
| 147. D | |
| 148. D | |
| 149. A | |
| 150. E | |
| 151. B | |
| 152. C | |
| 153. B | |
| 154. E | |
| 155. A | |
| 156. E | |
| 157. C | |
| 158. A | |
| 159. C | |
| 160. E | |
| 161. E | |

**True - False**

1. False - gametogenesis is the formation of sperm and eggs
2. True
3. True
4. False - somatic cells are diploid
5. True
6. True
7. True
8. True
9. True
10. False - Leydig cells secrete testosterone
11. False - Forty-six doubled chromosomes
12. True
13. True
14. True
15. False - LH, not FSH
16. False - The prostate, not the bulbourethral glands
17. True
18. True
19. False - estrogen, not progesterone
20. False - diploid, not haploid
21. True
22. False - follicle, not corpus luteum
23. True
24. True
25. False - FSH, not LH
26. True
27. True
28. True

**Fill in the Blank**

29. menopause
30. granulosa cells, primary follicle
31. atresia
32. follicular cells
33. myometrium, endometrium
34. upper 1/3 of oviduct
35. seminal vesicles
36. resolution
37. erectile tissue
38. impotence

39. gametes
40. somatic cells
41. autosomal chromosomes
42. cryptorchidism
43. Leydig cells, interstitial cells
44. a head, an acrosome, a midpiece, a tail
45. luteinizing hormone (LH), follicle-stimulating hormone (FSH)
46. blastocyst
47. decidua
48. chorion
49. hCG (Human chorionic gonadotropin)
50. 38
51. Oxytocin
52. involution
53. Colostrum

**Matching**
54. A,B,E
55. D
56. B
57. C
58. D
59. A

**Multiple Choice**
60. C
61. D
62. A
63. D
64. D
65. E
66. C
67. B
68. C
69. D